SpaceTEC®
National Aerospace Technical Education
Center

Championing Tomorrow's U.S. Technical Workforce:

A National Priority for Future Success

By
Albert M. Koller, Jr., Principal Investigator, SpaceTEC®, and
Executive Director of Aerospace Programs, Brevard
Community College, FL

Patricia A. Cunniff, Co-Principal Investigator, Prince George's
Community College, MD, Editor

This SpaceTEC® publication was prepared with support from the National
Science Foundation

SpaceTEC®
National Resource Center for Aerospace Technical Education
7099 N. Atlantic Avenue, Suite 300
Cape Canaveral, Florida 32920

Library of Congress Control Number 2016917578
ISBN 9780966857030
Copyright © 2004
Revised Copyright © 2016

Printed in the United States

Table of Contents

Prologue

Thomas Gamble
President, Brevard Community College

The Community Colleges for Innovative Technology Transfer (CCITT), which I serve as Vice President, is a national consortium of community colleges that was formed during the first year of the NSF ATE program (1994) to conduct a project to infuse NASA technology into the curriculum for the workplace of the future for geospatial information technology.

SpaceTEC® grew out of this same partnership, addressing the need to improve educational programs for this country's technical workforce in the aerospace industry. In January of 2000 a "Space Summit" was held at NASA's John F. Kennedy Space Center and attended by key NASA and Florida officials who identified issues with the aging workforce and the need for career development opportunities for technicians nationwide. These conclusions were reflected in similar recommendations by the President's Commission on the Future of the U. S. Aerospace Industry (the Walker Commission Report). As a result of these concerns, work began on development and delivery of educational programs designed to produce graduates capable of assuming the roles of veteran technicians and expert practitioners who have pioneered our country's space program over the past 40+ years.

As the president of the host college for SpaceTEC®, I am pleased to offer this monograph as a testimony to the effectiveness of the SpaceTEC® program and the important work represented here.

Thomas E. Gamble is President of Brevard Community College, Florida, and Vice President of the Community Colleges for Innovative Technology Transfer (CCITT).

The goals and activities of the partners in this project have been directed at industry-driven needs for qualified employees who can keep us globally competitive in leading-edge technologies. Since this project began, SpaceTEC® colleges have produced graduates meeting those needs, and the partnership continues to grow in numbers and in its ability to create programs that address emerging areas of advanced technology that are much broader than those exclusive to aerospace. The national reach of the program has inspired spin-off activities that enrich the participating colleges. Local partnerships with business and government have been strengthened and made more productive. Clearly, SpaceTEC® has been of benefit to all participants, but most significantly to the students who have registered for the programs and completed the degrees and certificates that lead to well-paying jobs and future academic success.

As SpaceTEC® enters its third year, even greater challenges and opportunities await us. The early assumptions about the skills and competencies required for success in the future workplace have already been validated. Industry partners have given SpaceTEC® a true national presence through a unique National Aerospace Technology Advisory Committee that functions to assess and endorse the curricula and certification processes of the project.

The National Science Foundation (NSF) provides support to community colleges in many forms. Some of the most effective projects have come from its Advanced Technological Education (ATE) Program. For the past 10 years the National Science Foundation has directed specific attention to developing new initiatives in science, technology, engineering, and mathematics through regional and national projects. At an ATE "Presidents' Conference" hosted last spring by NSF, it became clear that the gains outlined for this project are typical of the experiences of my colleagues and their institutions. We are grateful for the NSF ATE Program and believe it has made a significant contribution to our SpaceTEC® colleges and to all community colleges across the country.

Foreword

Rich Lake
President, American Technical Education Association

In 2003, the NSF SpaceTEC® National Aerospace Technical Education Center joined the American Technical Education Association (ATEA) as an affiliate member – the first in ATEA's 76 year history. ATEA is the nation's premiere association for the technical educator, with emphasis on professional development, and SpaceTEC® is the primary national resource for aerospace technical education. Both groups have collaborated to promote quality technical education and professional development activities for postsecondary faculty at their respective member institutions.

Activities to support professional development of faculty take many forms, but this partnership has been especially fruitful in finding new ways to promote interest and foster commitments to "up-skilling" faculty and professional staff. While SpaceTEC® has special expertise in aerospace technical education, ATEA has special expertise in regional and national conferencing for updating of technical educators. Bringing these groups together has allowed each to benefit from the other's expertise. It has helped avoid duplication of effort and prevented the offering of low quality programs.

Richard Lake
President
American Technical Education Association

Dean of Career Programs
Parkland College, Champaign, IL

Additional shared goals and interests include efficiencies of scale, interest in building and disseminating models of technical curriculum and delivery, preparing a well educated workforce for our nation, documentation of preparation that can be recognized nationwide, and advocacy for the importance of technical education in our country's workforce development system – indeed that the technical and community colleges in the United States constitute the very foundation and structure of our workforce development system.

This Monograph constitutes another indicator of the commitment of ATEA and SpaceTEC® to partnering for the improvement of technical education. It describes a model program developed for the aerospace industry that can be duplicated in a variety of other industries. It is a promising practice to strengthen the industrial base that is the foundation of the U.S. economy. ATEA and SpaceTEC® encourage you to learn about this model and its potential for improving math, science, engineering, and technical education in your own area.

Preface

Patricia A. Cunniff, Editor
Prince George's Community College

In this Monograph, we have attempted to tell the collective story of SpaceTEC®, our NSF National Center of Excellence, and to provide an individual perspective from each of our partner colleges. The result is far more extensive, and we hope more interesting, than the required final report of any funded project. Our SpaceTEC® colleges are located from coast to coast. Each institution is very different from the next. Likewise, our local industry base varies enormously. Despite this diversity, we have worked together to share curricula and then to tailor the aerospace technology core to local needs. We believe these experiences will be valuable to other colleges throughout the country which are grappling with meeting local industry needs while laboring within tight fiscal constraints.

One "Lesson Learned" that shows up in each college narrative is the need for strong partnerships with industry. Local industry provides needed input on curriculum development and validation, recruitment of students and adjunct faculty, assistance with skill assessment, resource development, and help with job placement of graduates. Building and nurturing these industrial partnerships will ensure that tomorrow's U.S. technical workforce is strong and competitive.

This Monograph represents the collective effort of many individuals. My thanks go to all the Co-PIs who provided the best about their colleges, to the SpaceTEC® National Center Staff which fostered this effort, and to the National Science Foundation which provided needed support.

Patricia A. Cunniff is Co-Principal Investigator, Prince George's Community College, Largo, MD, and Editor, SpaceTEC® Monograph.

Executive Summary

Albert Koller
Principal Investigator, SpaceTEC®

In 2002, the National Science Foundation (NSF) established SpaceTEC®, the National Aerospace Technical Education Center, to serve as a focal point for aerospace technology education for the future in response to the nation's aging technical workforce. Using formal partnerships with business and industry linked to the nation's

community college system, this project relies heavily upon industry-based requirements for identifying, teaching, and examining the skills of the technician of the future (see photo at left). SpaceTEC® now includes 12 colleges in eight states, including Embry-Riddle Aeronautical University. The many partnerships include a national advisory committee with representatives from the aerospace industry; government aerospace agencies from Alabama, California, Florida, Mississippi, Ohio, and Texas; and five NASA centers.

Since its inception. SpaceTEC® colleges have placed two-year Associate's Degree graduates in aerospace jobs; established regional and national advisory committees; and fostered articulation agreements with K-12 and higher education.

To date, SpaceTEC® colleges have graduated more than 250 technician trainees with Associate degrees; completed outreach workshops and professional development for more than 45,000 teachers, students, parents, and industry partners; and secured access to Launch Complex 47 at the Cape Canaveral Air Force Station (see photo at right) to support practice-based training activities.

According to industry estimates, the SpaceTEC® program will save employers a minimum of $10,000 to $12,000 per employee in orientation and training costs for each new hire with an Associate's Degree.

Over the short history of the project, SpaceTEC® colleges have added new facilities, curricula, and programs, including access to a small missile processing facility and an active launch complex on Cape Canaveral Air Force Station. Several colleges have completely revised their technical curriculum with industry participation; opened new programs or renewed existing programs in Engineering Technology, Space Technology, Composites, Metrology, and Geospatial Technology; placed their graduates in jobs at leading aerospace companies; established new K-12 initiatives including robotics demonstration programs and live downlinks from the International Space Station; and developed articulations with Embry-Riddle Aeronautical University for all SpaceTEC® colleges.

As important as these primary achievements are to the overall success of the program, perhaps the most significant lasting benefits will come from the additional opportunities available to all of the colleges because of the outstanding working relationships that have developed among the many partners. From the sharing of curricula and expertise to the joint development of skill standards and the strengthening of programs of mutual benefit, this consortium of institutions and their partners in business and government has produced a long list of notable achievements that go far "above and beyond" those promised in the grant.

The next step in the development process will be the expansion of current aerospace skills and competencies in the core curriculum to major elements of aerospace specialization beyond the national core. Over the next year, SpaceTEC® colleges will put in place this country's first-ever performance-based national certification examination for aerospace technicians using skill standards derived from industry and endorsed by the key organizations in all sectors. Application of these approaches to career-ladder professional development programs and other related areas of applied technology will follow. How far SpaceTEC® can move beyond that goal remains to be seen.

None of this work would have been possible without the strong support and financial backing of the National Science Foundation's Advanced Technological Education Program. We deeply appreciate the mentoring, creative thinking, and funding provided by the National Science Foundation and the ATE Program Officers who facilitated these accomplishments.

For those interested in a broader view of this project and its continuing work, details of SpaceTEC® activities can be found at: www.spacetec.org.

Albert M. Koller, Jr. is Principal Investigator of SpaceTEC®, the NSF ATE Program's National Center of Excellence for Aerospace Technical Education, and Executive Director of Aerospace Programs at the Brevard Community College in Florida.

Project Overview

David Brotemarkle
Program Manager, SpaceTEC®

The president's announcement of new goals for the nation's space program--including a return to the moon to establish a permanent lunar base, completion of the space station for near-earth research, and long term manned flights to Mars–is exciting news that has stimulated greater interest in all space-related programs. To address the need for skilled technicians and to spur student interest and motivation, a consortium of colleges called the **Community Colleges for Innovative Technology Transfer** (CCITT) has established **SpaceTEC®**, a national aerospace science Technical Education Center of Excellence funded by the NSF's Advanced Technological Education Program (ATE).

SpaceTEC® has grown from local programs at several community colleges to a national effort with broad industry support and an operational national infrastructure. Brevard Community College serves as the fiscal agent and Center manager, and there are now **twelve member colleges participating** in eight states. All are affiliated with a NASA center or Department of Defense location. Their activities for SpaceTEC® range from conducting clinics and workshops for K-12 teachers and students to offering Associate's Degrees and Advanced Technical Diplomas. Their primary focus is the development and delivery of hands-on technical education designed to meet employment needs and national skills standards being identified under the SpaceTEC® program to support a national technician certification exam program. Collectively their respective institutions enroll over 400,000 students annually. A map of the primary participants is shown on the next page.

From its inception, SpaceTEC® has operated as a dispersed Center with delegation for deliverables to each of the college partners.

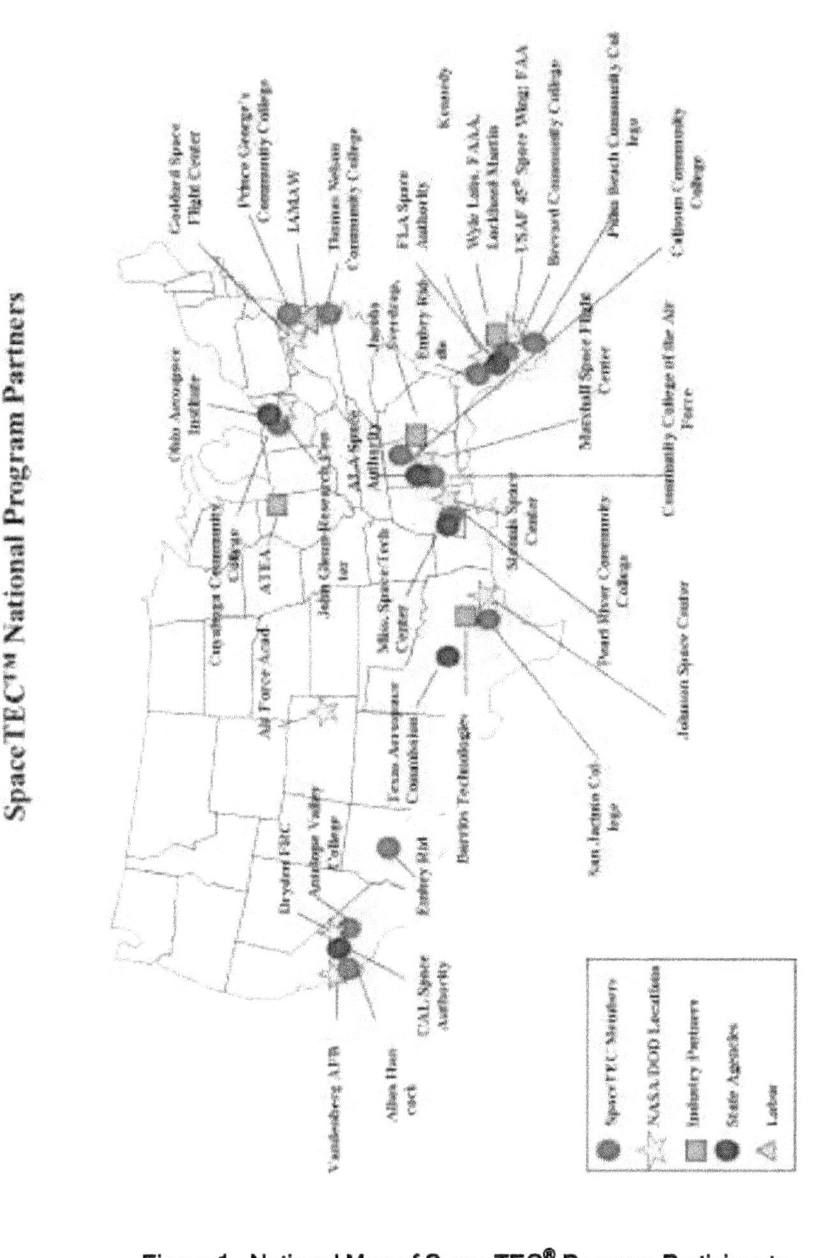

Figure 1. National Map of SpaceTEC® Program Participants.

Industry support has been an important part of the SpaceTEC® project. Jacobs Sverdrup and United Space Alliance have provided industry Co-PIs, and partner colleges lead one or more of the elements of the national program.

One of the lessons learned in SpaceTEC® management is the need for regular update and consistent, formal reporting of all major work processes for a decentralized Center relying on its partner colleges for the primary deliverables. SpaceTEC® has used a combination of monthly teleconferences, written monthly reports, quarterly newsletters, an active website, and quarterly Co-PI meetings to maintain open communications that assure the sharing of information, updates, and status on all current operations. Sub-award contracts have been used to detail tasks, budgets and schedules for all participants.

SpaceTEC National Center of Excellence
NSF Advanced Technology Education Program

Nat'l Visiting Committee	Executive Director	National ATAC
Dr. Robert Sullivan, Chair	Dr. Albert Koller, PI	Marshall Heard, Chair

External Evaluator	Internal Evaluators
Col. David Bossert, USAF	Dr. R. Hinkle, Dynamac G. Hauer, Wyle Labs

SpaceTEC Coordinator	Program Manager	Technical Support
J. Curtis	D. Brotemarkle	S. Pantano

Faculty Development	K-12 Programs	Marketing/PR	Databases
Dr. P. Cunniff, PGCC Dr. M. Dalton, SJCC	Dr. T. Steffen, PBCC E. Smith, PRCC	J. Swindell, Calhoun J. Martin, Calhoun	A. Neilson, AHC Fred Johnson, AHC

Certification Exams	Curriculum	Special Projects	New Initiatives
R. Lengyel, Consultant G. Grimshaw, DFRC F. Longo, USA	J. Swindell, Calhoun G. Strohm, BCC JR Breeding, CCAF	M. Drake, AVC Dr. C. George, TH-C Dr. P. Cunniff, PGCC	Dr. P. Taylor, TNCC Dr. D. Hosley, ERAU D. Welsh, PRCC

Figure 2. SpaceTEC® Organizational Structure.

To meet the stringent NSF requirements of a national center, policy guidance, advice, and project oversight are provided by four groups:

(1) A National Visiting Committee, composed of ten volunteers selected by the NSF from senior leaders, meeting annually to review the Center's operations and offering observations and recommendations to the National Science Foundation.

(2) The National Aerospace Technology Advisory Committee (NATAC), meeting at the call of the Chair and operating through a formal committee system to review and endorse the major elements of the project, providing the national oversight for industry and government.

(3) Internal evaluation by **Dr. Ross Hinkle, Dynamac Corporation** and **Mr. George Hauer, Wyle Laboratories**.

(4) External Evaluation by **Lt. Col. David Bossert of the Air Force Academy**, providing both summative and formative evaluations of project operations.

The listings of the members of these groups are shown at the back of this Monograph.

A national infrastructure system has been created to link the local activities of community colleges with the national scope of the aerospace industry across government agencies and non-profit groups as shown in Figure 3. This system is essential to successful development and implementation of a national skills standard for aerospace technicians, and in August 2003 this schema was activated and all elements are operational.

This infrastructure provides, for the first time, professional career development activities for aerospace technicians, offering a national curriculum, contact and activity databases, job banks for both employers and prospective employees, methods for networking and maintaining contact with colleagues and potential team members such as a web site, newsletters, conferences, and journals, and access to testing for competency in national skill sets that can be verified for knowledge and performance to qualify candidates entering employment for the first time or "up-skilling" for future advancement of incumbent workers.

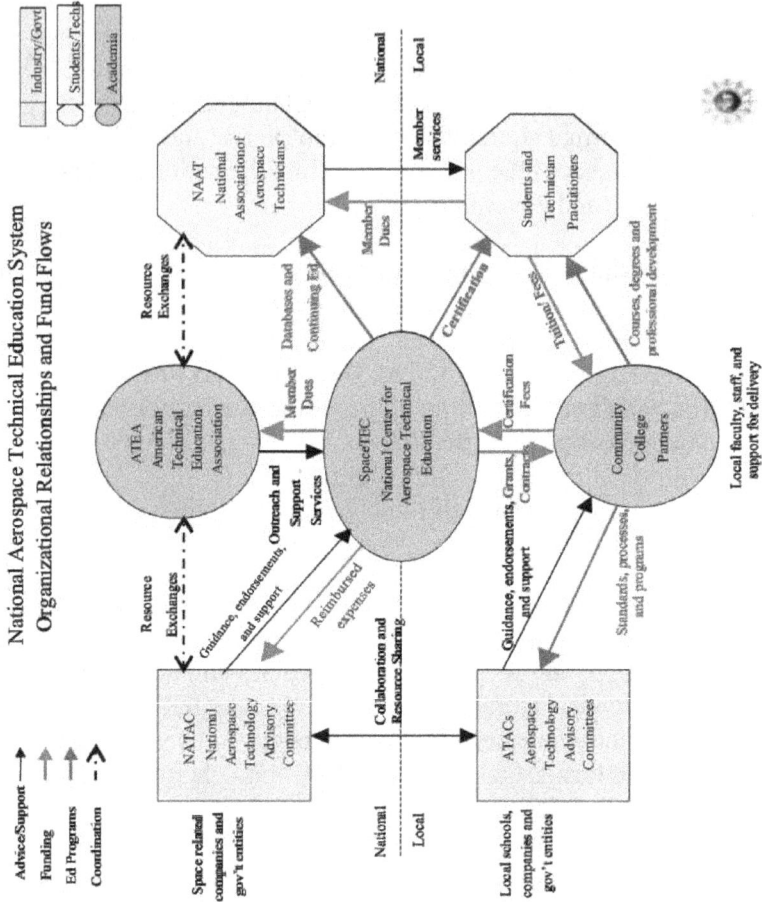

National Aerospace Technical Education System
Organizational Relationships and Fund Flows

Figure 3. SpaceTEC® National Technical Education Infrastructure.

Depicted on the left side are the relationships between the local and national advisory committees, providing guidance and support to assure the quality of the curriculum as well as access to workshops, labs, and subject matter experts from the workplace.   In the center are symbols for the education sector: community colleges – the entry point for all educational and testing activities -- with linkages to SpaceTEC® as the certifying agency, and the ATEA as the premiere agency providing access to technical faculty nationwide for professional development. On the right hand side of the chart are the National Association of Aerospace Technicians and the students and practitioners who are the primary focal points for this project.

The national infrastructure is an integral part of the certification program being deployed during Year 3 activities. These include the development of the examination process, the qualification, training, and certification of examiners who will administer the examination at various locations, and the creation of examination support systems suitable for implementing a system that can be successfully replicated wherever needed.

The examination process has been developed using several applicable models, including the Federal Aviation Administrator's examination for Airframe and Powerplant Mechanics. The SpaceTEC® examination is in two distinct parts—a core examination that is based on the national core elements of the curriculum, and a series of concentration examinations that are based on specialty training that is customized for each locale.

The core examination has been developed with inputs from all of the SpaceTEC® partner colleges and consists of a computer-based, Internet accessible job knowledge examination and a laboratory-based hands-on performance examination demonstrating workplace competencies required of aerospace technicians. The concentration examinations will follow a similar approach, including a specialized knowledge examination and a competency-based performance examination to be administered by a certified SpaceTEC® examiner.

The benefits of SpaceTEC® are many. Program **graduates** receive nationally recognized competencies for employability. Participating **academic institutions** receive funding for their roles, access to curriculum, program articulation, faculty development, and recognition for industry-endorsed programs for critical workforce needs. **Business and industry** are provided with a dependable source of well-educated entry-level technicians and a means of sustaining workforce development. **Government** is provided an improved educational capability and qualified technicians for a key economic sector. **Educators** have new space-related curricular themes to enrich their courses in many disciplines.

The sections that follow describe the approaches, accomplishments, and lessons learned by each of the partner colleges and their contributions to the national system that has emerged.

David Brotemarkle is the SpaceTEC® Program Manager and a retired USAF colonel and command pilot.

Partner Profiles
and
Perspectives

Allan Hancock College
Santa Maria, CA

Background

Allan Hancock College (AHC) is an accredited, two-year, post-secondary, California community college. Founded in 1920, college classes were initially offered at Santa Maria High School and were later moved in 1957 to the Hancock Field. The college has a rich aeronautical historical connection since Hancock Field previously housed the Hancock College of Aeronautics, followed by the University of Southern California's School of Aeronautics. The college was later renamed after Captain G. Allan Hancock, a prominent citizen who owned Hancock airfield and adjacent land.

Allan Hancock College, a federally designated Hispanic serving institution, enrolls 12,000 credit students each semester from northern Santa Barbara County and parts of San Luis Obispo and Ventura Counties. More than 1,300 courses are offered in over 100 fields. Over 5,000 students take noncredit courses each semester.

AHC's main campus is located in Santa Maria. Additional Centers are located in Lompoc, Solvang and at Vandenberg AFB. Aerospace, government, and agriculture are major employers in the region. Vandenberg AFB is the second largest employer in Santa Barbara County. Allan Hancock College, with more than 1,000 employees, is one of the 12 largest employers.

In addition to general education, the college offers programs in a number of highly specialized fields. In fall 2004, a new A.S. degree and certificate in electronics technology–space operations will join the list of specialized offerings.

Participation
Allan Hancock College has undertaken the following to support the SpaceTEC® goals:

- Develop and maintain databases for grant partners and public use in curriculum, contact/membership, and assessment and certification areas. These databases will expedite sharing information and data research and archiving for internal and external use.
- Develop a new associate of science degree and a certificate in alignment with grant core curriculum focusing on a concentration unique to our college, electronics technology - space operations. Graduates from the program will be well trained and available for local positions in the aerospace industry at Vandenberg AFB, Edwards AFB, and surrounding Southern California areas.
- Participate in student recruitment for the new academic program and outreach to attract new students and make them aware of career opportunities for aerospace technicians.
- Offer a K-12 teacher outreach activity to educate and excite faculty about space and science-related curricula.
- Complete articulation agreements with four-year colleges, e.g., Embry-Riddle Aeronautical University, California Polytechnic State University, San Luis Obispo and Pomona campuses. This will help establish educational pathways for transfer students who wish to complete a four-year degree.
- Become an aerospace technician certification test site and help coordinate training and site activities. This will further help to establish the importance and credibility of the skills and knowledge level of aerospace technicians.
- Host an NSF SpaceTEC® national grant partner meeting to share information, discuss issues, develop future plans, and showcase the host college's on- and off-campus sites that relate to aerospace, e.g., tours of aerospace corporations and Vandenberg AFB training/education and operations sites.

- Develop a local advisory group to help provide input for curriculum development, student recruitment, and internships.

The development of databases focused on grant goals and was refined as partners provided input. Ongoing maintenance issues such as security and firewall protection continue. The databases will continue to grow in complexity and size.

The new associate of science degree and certificate, electronics technology - space operations, evolved from local, state, and national input such as the college president's executive aerospace roundtable luncheon with over 20 industry and community leaders in attendance; industry input from advisory meetings, and state conferences and regional meetings; and the national input from the grant partners. Curriculum was obtained from grant partners to minimize time spent in course development and take advantage of what was already working well at other colleges. The curriculum was developed with not only the needs of the workforce in mind, but also the local student population's possible barriers to enrolling in classes, i.e. introducing more online and open lab opportunities to support those who work full time or who are single parents.

It continues to be important to collaborate with industry, education, and other nonprofit agencies on curriculum development and student/faculty outreach. California Space Authority, Lockheed Martin Corporation, Boeing Company, and ITT Industries are important partners. On campus, we work closely with the MESA program to help students learn about aerospace job opportunities.

Results
Curriculum/Degree/Certificate
Three new space-oriented classes were adapted from courses made available from the National Science Foundation SpaceTEC® consortium – two from Brevard Community College, Florida, and one from Capitol College, Maryland. (* below indicates the courses) and approved as new courses at AHC. Space 102 will be offered fall semester 2004. Space 104 and 128 will be introduced in spring 2005.

List of new courses created and approved:
*Space 102   Introduction to Space
*Space 104   Quality Management Control and Safety
*Space 128   Materials and Processes
 EL 179   Workshop (Open Access Lab)
 Space 179   Educators' Launch Workshop (Gravity Probe
        B Launch: NASA's Relativity Mission, Testing
        Einstein's Universe)

Databases

All databases initially specified in the grant have been developed and are operational. Usernames and password management is in place. Additional databases have been created to help support grant activities, i.e. calendar database and job postings database. The database team has taken on the additional responsibility of developing and implementing a new web-based operational system. This system was donated by the lead programmer to the grant and is being adapted to fit the grant's needs.

Outreach Activities

- **Educators' Launch Conference**

National publicity about the $700 million *Gravity Probe B Launch: NASA's Relativity Mission, Testing Einstein's Universe*, helped to excite teachers attending the spring 2004 K-12 outreach activity at Allan Hancock College's Lompoc Valley Center. The Educators' Launch Conference focused on new curriculum ideas to help teachers understand the launch and the research on which it was based, so they could present the space science curriculum in K – 12 classrooms. SpaceTEC® was one of many sponsors of this very successful activity.

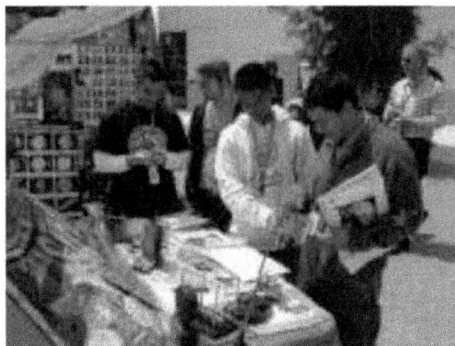

Outreach activities at Allan Hancock support SpaceTEC® goals.

- ***Agriculture & Industrial Technology Expo & Job Fair***

The NSF SpaceTEC® booth was the most attended exhibit at the Allan Hancock College *Agriculture & Industrial Technology Expo & Job Fair,* a major outreach event sponsored by the college and held April 1, 2004. Our SpaceTEC® team manned the booth providing information about the grant and our new electronics space operations A.S. degree/certificate. Over 480 junior and senior high school students along with several hundred members of the general public, and college staff and students attended the Job Fair.

- **Women in Engineering Speaker Panel**

SpaceTEC® co-sponsored with the MESA program a guest panel of women professionals in various aerospace engineering-related careers. Over 100 persons attended the panel presentation featuring mechanical and electrical engineers and VAFB employees.

Allan Hancock student works on circuit board.

Lessons Learned

Networking and industry partnering are crucial to success. The synergy grows in proportion to the networking and partnering efforts. One must step outside the classroom or office and into the community to understand employers' needs, obtain industry donations, help with student and faculty recruitment, and find opportunities to share facilities, equipment, and curricula. For example, the latest articulation agreement with Embry–Riddle Aeronautical University at Vandenberg AFB, is helping us promote the new electronics technology–space operations A.S. degree. Dr. Larry Gooch, our national representative on the SpaceTEC®

advisory committee, who is also a retired USAF commander and member of the California Space Authority, is helping our college secure equipment donations from VAFB aerospace corporations.

Even though we have cited massive retirements and foreign hiring by local aerospace corporations, more documentation is needed to demonstrate to the college community and the community at large that the aerospace industry will soon face serious labor shortages, especially in specific technical areas. A major public relations campaign is needed nationally to deal with projected labor needs so as to build future college enrollments in appropriate academic areas.

SpaceTEC® has been a gift to our college, helping to revitalize a curriculum, bring new cutting edge curricula to the forefront, update equipment through leveraged funds, and re-inspire faculty and administrators alike. The national input has provided a depth to our discussions that could not take place without the meetings and dialogues provided by the grant. New ideas have emerged, changed, evolved and come to fruition.

Future Activities
The Allan Hancock SpaceTEC® team will continue to strengthen relationships with Vandenberg AFB and local aerospace corporations to enrich and develop the new aerospace electronics technician program and develop new workforce based curricula. A new degree, Mechatronics, is being developed to meet local and national aerospace workforce needs. Curriculum development is supported by a NASA Curriculum Improvement Partnership Award. More outreach events, on- and off-campus, will be planned to attract media attention and recruit more students into electronics.

College District President/Superintendent
Dr. Ann Foxworthy
Allan Hancock College
800 South College Drive
Santa Maria CA 93454-6399

Co-PI/Contact
Ardis Neilsen
Interim Director
Economic Development
805-922-6966 x3325
aneilsen@hancockcollege.edu

Associate Superintendent/
Vice President Academic Affairs
Dr. Betty Inclan
Dean, Academic Affairs
Dr. Ray Hobson

Team Members
Joyce Krantz
Fred Patrick
Robert Alldredge,
Judy Markline, Dean

Antelope Valley College
Lancaster, CA

Background

Antelope Valley College, located in the high desert of southern California, serves a 1,945 square mile region as the area's only college. Associate Degree and Certificate Programs are offered in more than 40 subject areas. The college was founded in 1929 and serves an estimated 13,000 students. Sixty-three percent of the students are female and forty-six percent are minority. Our main campus is located on 125 acres in Lancaster and some aviation courses are taught at Fox Airfield, Lancaster.

The local community is very active in aerospace activities. Northrop Grumman is the largest employer, and Mojave is home to Scaled Composites—the country's first private commercial space venture.

Participation

We are very proud of our Aviation Airframe and Powerplant training programs. While participating in SpaceTEC® we have initiated a Certificate Program in Aviation Composites Fabrication and an AS Degree in Aviation Fabrication and Assembly. Antelope Valley College also offers a program in Air Traffic Control.

Antelope Valley joined SpaceTec® one year after the grant was funded. The college has developed a state-of-the-art curriculum in aircraft fabrication and laboratory facilities for aerospace composites. AVC also conducted a SpaceTEC® faculty workshop in composites, summer 2004, and shared course documents and laboratory plans with interested partners.

Workshop participants visited SpaceShipOne at Scaled Composites in the Mojave Desert.

Results

Antelope Valley College hosted the first training for SpaceTEC®
Examiners (STE) on the West Coast and is now prepared to offer
the SpaceTEC® Core Certification Exam at its Mojave facility.

Hands-on training has included working with composite
materials, blueprint reading and learning computer software
programs. Among the beneficiaries have been such companies as
Lockheed Martin, Crissair, Fiberset, British Aerospace Systems,
Boeing, Plant 42 and Edwards Air Force Base, and Northrop
Grumman.

The AVC Airframe and Powerplant Program has been
strengthened by the SpaceTEC® sponsored needs assessment
and broadening of the partnering network inherent in this
project.

In June 2004, SpaceTEC® Composites Workshop faculty toured the
Scaled Composites facility to see the White Knight turbojet aircraft.

Lessons Learned

Antelope Valley College is perfectly located amid numerous aerospace contractor facilities. Networking and partnerships are our key to success. Retired aerospace workers serving as adjunct faculty enrich our curriculum and establish needed communications within companies.

SpaceTEC® participation has facilitated training in new test equipment and the development and activation of new training laboratories. Partner schools have supplied original K-12 materials in space themes for use in local systems. Participation in SpaceTEC® provides many unique opportunities and contacts.

The SpaceTEC® grant complements several other initiatives at Antelope Valley, specifically the $4M (state and federal monies) to provide skills training for existing or potential employees in the industry, and the $2.8M training grant from the Department of Labor. The latter has the distinction of being the only federal program in California training incumbent workers.

Future Activities

AVC will support the SpaceTEC® certification program as a testing center for the national core certification exam. At least two certified SpaceTEC® Examiners (STE) will be available to use the SpaceTEC® Examination Kit housed in college laboratories.

AVC will offer a second (and perhaps subsequent) composites workshop for SpaceTEC® partner college faculty.

AVC will continue to support SpaceTEC® activities, including outreach to local K-12 schools and curriculum development in emerging aerospace technologies.

Participants at the SpaceTEC® Examiners Training held June 2004 at Antelope Valley College.

President
Dr. Jackie L. Fisher, Sr.
Antelope Valley College
3041 West Avenue K
Lancaster, CA 93536-5426
661-722-6300

Co-PI/Contact
Margaret Drake
Dean, Technical Education
Antelope Valley College
3041 West Avenue K
Lancaster, CA 93536-5426
661-722-6327
mdrake@avc.edu

Dean of Technical Education
Margaret Drake

Team Members
Marie Clinton
John Petersen

Brevard Community College
Aerospace Program
Kennedy Space Center, FL

Background
Brevard Community College's Aerospace Program is headquartered
at the Spaceport Center which is located in the Center for Space
Education at Kennedy Space Center, Florida. Brevard Community
College also serves as host to the National Science Foundation's
SpaceTEC®, which is a National Center of Excellence located at
Cape Canaveral Air Force Station.

The Aerospace Technology Program is a two-year Associate of
Applied Science/Associate of Science degree program that consists
of seventy (70) credit hours. The Aerospace Program prepares
students for employment as aerospace technicians that manufacture,

K-12 Teachers at Brevard Community College Rocket Workshop.

assemble, service, test, operate, maintain and repair systems associated with expendable and reusable space launch vehicles, payloads, related laboratories and ground support equipment.

The Aerospace Program, established in 2001, began enrolling students in August 2001. The program has grown rapidly in a short time into four full-time cohort groups attending classes with an average of sixty (60) students enrolled each semester. The "First Class" graduated in May 2003 and two additional classes have since graduated. The majority of graduates have obtained employment through United Space Alliance, which is great supporter of the program.

"Gemini" class graduates with Kennedy Space Center Director, Mr. James Kennedy.

The program continues to flourish as we expand our partnerships into the local area. Our Aerospace Technology Advisory Committee consists of nearly 40 members from industry, government, and academia. This group plays a key role in student and faculty development, curriculum decisions to keep the program current with local industry needs, internships, mentoring, and articulation of the two-year degree to four-year programs.

The Aerospace Program has also expanded its reach into the neighboring school district. Providing local technology, math and science teachers with space-related resources, classroom visits for rocket workshops, job fairs, career days, and teacher aerospace and rocket workshops highlight the program's commitment to bring "Space to Schools" and "Schools to Space."

Participation
Initially a training and education need was identified based on changes to the workforce and the way the aerospace industry operated. As the workforce aged and retired, there were too few trained people available. Industry could no longer financially support the extensive training needed to bring inexperienced workers to the competency required. The plan was to coordinate the efforts of local industry, government, and academia to find solutions to present and future workforce need. The Aerospace Technology Advisory Committee (ATAC) was formed to fulfill this role.

Key members of ATAC worked to develop the curriculum that would encapsulate and outline the requirements for entry-level workers. Once this baseline was developed and endorsed, Brevard Community College's Aerospace Technology Degree Program courses began at the Center for Space Education in fall 2001.

As a result of the curriculum collaboration, students will learn about the history of space technology and spaceflight, propulsion and orbital science, lunar and planetary exploration, Earth-orbiting satellites, regulatory issues, launch and payload processing activities, electronic fabrication, fiber optics, fluid systems and fabrication of metals, and composites. Course activities will provide hands-on instruction in the use of tools, equipment, materials, and current practices and processes found in the industry.

Graduates of the program will be qualified for entry-level positions in the aerospace industry, with a standardized and industry-endorsed education program that provides employers with a well-trained and productive workforce on both reusable and expendable space vehicles. Graduates may also qualify for many applied-technology jobs such as testing, fabrication, assembly, repair, and manufacturing.

Instruction is designed to qualify students for technical skill assessment evaluations and immediate employment. This program also provides supplemental training for persons previously or currently employed in this occupation.

Students working on "composites" project.

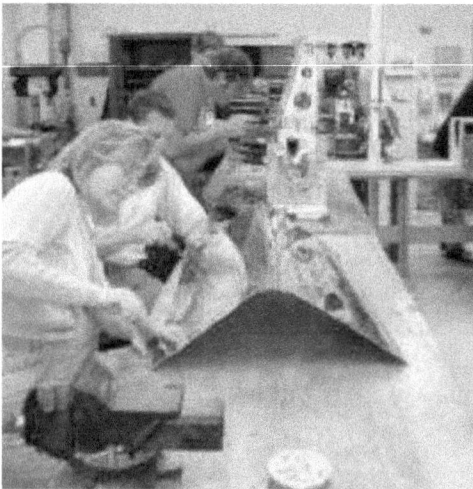

Students working on sheet metal fabrication "winglets."

The approach was to combine the efforts of the partnership formed by the ATAC to provide solutions to local workforce issues. Key was the industry's commitment to provide dedicated and highly trained personnel to deliver the course content and share relevant work experiences with the new degree-seeking students. Once the curriculum was developed, endorsed, and the delivery process began, there was a need to expand the breadth of the program to include the following:

- Determine Core and Concentration area curriculum standards.
- Develop national certification program.
- Articulate 2-year programs to 4-year programs.
- Recruit students at the local and national levels.
- Develop outreach activities to include K-12 students and teachers.
- Provide co-op, temporary and internship opportunities for students.

Results

- Obtained articulation agreements with Embry-Riddle Aeronautical University.
- Graduated three classes of students as of May 2004.
- Obtained cooperative and temporary work positions for students enrolled in degree program.
- Expanded outreach opportunities into K-12 for students and teachers to include:
    o Classroom visits
    o Job Fairs and Career Days
    o Teacher Workshops
    o Summer Industrial Fellowship for Teachers (SIFT)
    o Civil Air Patrol links
    o Resource and Curriculum Development
- Successfully operated summer internship program.
- Developed online Ethics course.
- Conducted first ever aerospace dual enrollment program.
- Coordinated job placement training with local JobLink.
- Organized and maintained local ATAC.

Lessons Learned
- Recruiting is very sensitive to industry changes--job placement drives recruiting potential.
- Developing strong K-12 linkages is very challenging--finding the key people is critical.
- "Space" still has a strong attraction to most people--young and old alike.
- Partnerships are the key to success. "Buy-in" from industry, government, and academe ensure support for curriculum development, recruitment of students/faculty, industrial tours and materials, and strong communication and dissemination of program information.

Future Activities
The team will continue to build on the successes of the past by seeking to expand the role of technology in K-12 education, and by searching for more partnership opportunities to diversify the role technology training plays in the development of students and in the workforce. In addition, we expect to:
- Explore additional articulation opportunities.
- Develop certification for curriculum concentration area.
- Develop additional online classroom opportunities.
- Expand dual enrollment programs into high schools.
- Develop linkages with additional employers for expanded placement opportunities.

College District President
Dr. Thomas E. Gamble
1519 Clearlake Road
Cocoa, FL 32922

**Executive Vice President/
Chief Learning Officer**
Dr. Donald P. Astrab
1519 Clearlake Road
Cocoa, FL 32922

Executive Director, Aerospace Program
Dr. Al Koller
Mail Code BCC/SpaceTEC
Kennedy Space Center, FL 32899

Co-PI/Contact
George Daniel Strohm
BCC-A, M6-306 Rm 2000
Kennedy Space Center, FL, 32899
321-449-5001
strohmg@brevardcc.edu

Team Members
Peri Baker-Horner
Gregory Bickford
Cheryl Duff
Amanda Reed

Calhoun Community College
Decatur, AL

Background

Calhoun Community College is the largest community college in the Alabama College System with approximately 9,000 students. In the late 1990s, Calhoun began an integrated initiative with local business and industry to reassess the alignment of technical education programs with local needs. The need for advanced technology education was a clear outcome of this study related to the high concentration of high technology and manufacturing industry coupled with NASA and defense-related initiatives in the Huntsville-Decatur metro complex. This became even more focused in 1998 as The Boeing Company selected Decatur for the Delta IV Rocket factory. Calhoun Community College would serve as training site for both incumbent and new hires to the Delta facility and provide a future source of prospective employees for the regional aerospace industry through an Aerospace Technician degree program. After working with Boeing, NASA, and numerous other aerospace industry representatives, the state of Alabama approved an Associate of Applied Technology program in Aerospace Technology at Calhoun in May 2000.

Calhoun 's Aerospace Training Center.

Participation

In the Huntsville-Decatur metroplex, much of the NASA and aerospace industry focus is centered on manufacturing and the related electrical and mechanical testing. Consequently, Calhoun's aerospace technician program has a manufacturing focus with four separate concentrations: (1) machining and fabrication, (2) structures and assembly, (3) welding, and (4) electronics. While these specialties sound traditional in name, they extend into current state-of-the-art technologies such as 5-axis CNC machining, 3-D modeling, brake forming, work with composite materials, orbital tube and specialized material welding techniques, and advanced electronics manufacturing processes to complement a broad range of electronic topics.

CNC Skin milling operation in progress.

One of the major challenges with this program is limited resource materials such as textbooks and commercial training curricula that correlate directly with the course learning objectives. Many times, applicable materials are written for baccalaureate or higher level engineering students or are contained in government or industry standards. Putting this information in the context of a practical, application-based, hands-on approach to learning consistent with a technician level of training has required significant effort. The absence of existing national industry standards for the aerospace technician was also an area of focus in the grant, and a common denominator of core skills was needed to proceed with any type of skills certification or articulation between participating colleges.

Calhoun Community College has been a primary participant in the SpaceTEC® initiative since its inception. In addition, Calhoun was a participant in the first CCITT NSF grant funded in 1994. Within the SpaceTEC® grant, Calhoun has had primary responsibilities for:

Student Recruitment and Outreach
- Work with Brevard Community College and others to develop, pilot, and implement recruiting initiatives for the Aerospace and Advanced Technology programs at Calhoun Community College.
- Implement retention and outreach pathways through articulation, dual enrollment, or similar outreach pathways.

Develop and Deliver National Technical Education Program
- Work with other SpaceTEC® colleges to identify a core curriculum and develop a cross reference of skills and competencies between Calhoun's Aerospace Technology courses and the SpaceTEC® core.
- Support both local and national Aerospace Technology Advisory Committees (ATAC).
- Establish and maintain a national recruitment website and newsletter.
- Establish articulation agreements.
- Establish pathways of educational advancement from secondary to post-secondary to the university and beyond.

National Skill Assessments
- Work with other SpaceTEC® colleges in the identification of Aerospace Technician skill standards and ensure that these are correlated with Calhoun's Aerospace Technology course outcomes and learning objectives.
- Pilot core certification at Calhoun Community College.

Faculty Professional Development
- Participate in faculty workshops.
- Disseminate materials on aerospace disciplines to K-12 faculty.

Results
Student Recruitment and Outreach
- Developed a plan and calendar for recruiting and outreach with Calhoun's Aerospace and Advanced Technology programs. For year 1 and 2, these plans have included presentations which reached about 1,900 students in 16 area high schools, presenting program information at high school college nights, and hosting an annual Workforce Expo which includes industry and college presentation of career and program opportunities along with open forum discussions with industry representatives and motivational speakers.
- Presented information on career opportunities in the aerospace industry to high school counselors and K-12 educators as well as information modules for classroom use.
- Strengthened retention and outreach through secondary articulation of machining and welding courses. A new dual enrollment program in Alabama's education system entitled Early College Enrollment Program (ECEP) has been developed which allows qualifying high school juniors and seniors to enroll full time in career technical fields at Calhoun. Piloting will begin during school year 2004-2005.

Develop and Deliver National Technical Education Program
- Calhoun participated with area industry in performing a DACUM in Aerospace Manufacturing during Year 1.
- Calhoun maintained an active local Aerospace Technology Advisory Committee (ATAC) and provided two representatives for the National ATAC.
- Calhoun worked with Brevard Community College, Community College of the Air Force, and others in SpaceTEC® during Year 1 of the grant to establish common core skills for Aerospace Technology programs.
- In Year 1, Calhoun developed a logo for the SpaceTEC® initiative, produced an information brochure, and established a national recruitment website (www.SpaceTEC.org) with contact information and links to each participating college along with state and national publications and media information. *SpaceTEC® Talk* was formatted and published.

- Calhoun established an articulation with Athens State University's B.S. in Management of Technology and Applied Technology programs in Year 1 of the grant. An articulation with Embry-Riddle Aeronautical University's B.S .in Management of Technology has been drafted, signed by Calhoun, and submitted to ERAU for signature in Year 2.

SpaceTEC® Talk Newsletter

Calhoun Aerospace student.

National Skill Assessments

- In Year 2, Calhoun reformatted the structure of its Aerospace Technology program after conducting detailed reviews of all 33 program courses in Aerospace Technology with industry subject matter experts. The main outcome was to provide more cross discipline training regardless of specialization.
- This review validated curriculum content with SpaceTEC® core competencies and industry expectations to ensure that skill standards are properly imbedded in the curriculum.
- Consistent with the intent of providing courses and instruction leading to industry certifications, Calhoun developed a format for course instructor guides that will ensure consistency in content and appropriate correlation between instructional materials and related skill standards.

Faculty Professional Development
- Calhoun participated in SpaceTEC® workshops at San Jacinto, Prince George's, and Antelope Valley.
- Calhoun faculty presented information on SpaceTEC® to K-12 teachers at the 2003 Alabama Aerospace Instructors Assn. and at the 2004 AAIA conference (with San Jacinto).

Lessons Learned
- Build programs based on defined industry needs and stay connected with industry.
- Ensure that programs and courses have instructor guides and classroom/lab materials to facilitate learning outcomes.
- Communicate with different sectors and populations in many ways in order to ensure adequate student enrollment.

Future Activities
Instructor Guides
- Develop instructor guides for all Aerospace Technology core courses including review by industry experts.

Manufacturing Concentration Certification
- The SpaceTEC® approach to developing the core curriculum for certification will be applied to development of the aerospace manufacturing certification program in Year 3.

Recruitment
- Recruitment continues to be a challenge. Enrollment and graduation are key to program viability. A 12-month integrated approach to recruiting has been initiated identifying all student sectors, defining what drives decisions, who influences prospective students, and how to communicate most effectively.

Calhoun Community College President
Dr. Marilyn Beck
P.O. Box 2216
Decatur, AL 35609-2216

**Vice President of Instruction
And Student Services**
Dr. Theresa Hamilton

Co-PI/Contact
Jim Swindel
Asst. Dean of Technology Ed.
256-306-2539
jes@calhoun.edu

Team Members
Janet Martin
Sherman Banks
Mike Bridges
Dennis Holmes

Community College of the Air Force
Maxwell Air Force Base, AL

Background

The Community College of the Air Force (CCAF) was established in 1972 and serves the U.S. Air Force and Department of Defense by helping enlisted personnel obtain an accredited Associate in Applied Science (AAS) Degree. Authority to award the AAS degree was provided in Public Law 94-361, signed by President Gerald R. Ford on 14 July 1976.

CCAF is a multi-campus, federally chartered institution and confers the AAS degree as part of the Air University (AU) system through its regional accreditation with the Southern Association of Colleges and Schools Commission on Colleges (SACS/COC). The college awards the AAS degree designed for a specific Air Force occupation or specialty and awarded its first degree in April 1977. CCAF serves over 380,000 students enrolled in more than 120 technical training schools and 66 technical degree programs. This makes CCAF the largest multi-campus community college in the world.

The college's administrative center is located at Maxwell AFB, AL and consists of over 7,500 faculty members and education service offices located worldwide. The major technical training campuses are located at Keesler AFB, MS; Lackland AFB, TX; Goodfellow AFB, TX; Sheppard AFB, TX; Vandenberg AFB, CA; and the School of Aerospace Medicine, Brooks AFB, TX.

Community College of the Air Force Administrative Center, Maxwell AFB, AL

Specialized schools provide additional technical training and
Professional Military Education campuses provide the majority of
the leadership, management and military studies coursework
required for the CCAF degree. CCAF's Missile and Space Systems
Maintenance and Electronic Systems Technology degree programs
have over 1,800 space and missile-related students enrolled and
produced over 1,300 graduates. The Air Force currently has over
2,700 technicians serving in the missile and space systems
maintenance and electronics occupational specialties.

Participation

The Air Force missile and space systems maintenance and
electronics curriculum is diverse and spans across several highly
concentrated spectrums equivalent to SpaceTEC® requirements.
The formal education and structured on-the-job training processes
provide the skills and qualifications necessary for missile and space
systems maintenance and electronics technicians to: acquire,
assemble, transport, install, inspect, maintain, modify and launch
ground and air missiles and subsystems, spacelift boosters, payload
guidance and control systems, satellites and subsystems, and
environmental blast doors and valves; launch, control, track and
recover unmanned air vehicles (UAV) and related equipment;

assemble, operate,
fabricate, install, calibrate,
test and troubleshoot
specialized research and
development (R&D)
systems and subsystems;
repair, calibrate, modify
and manage related
missile, booster, satellite,
UAV and R&D facilities,
support systems, test
equipment and subsystems; monitor, analyze and compile system
performance data; and maintain automated and manual electronic
test, launch control, checkout and support equipment.

The U.S. Air Force does not solicit or use National Science
Foundation (NSF) grant funding. CCAF has received no monies for

its participation in the SpaceTEC® grant. However, the Air Force partnership with SpaceTEC® provides an excellent opportunity for Air Force missile and space systems maintenance technicians to pursue a nationally recognized professional certification in which they qualify. This opportunity only exists through the collective sharing of degree plan templates, requirements and curriculum. This partnership also shares the same goal of the certification process. CCAF's experience in certification program development coupled with the SpaceTEC® collective vision and goals resulted in streamlining the certification process itself.

Results
The U.S. Air Force and CCAF have been primary players in the SpaceTEC® initiative from its inception. CCAF's long-standing and comprehensive Missile and Space Maintenance and Electronics Systems Technology degree programs, coupled with the Air Force's role in space force employment, research and development, and technologies, have been significant contributing factors to SpaceTEC®. CCAF staff members have:
- Played a coordination role in establishing a national Aerospace Technology Advisory Committee.
- Served on the NSF National Visiting Committee.
- Assisted in the development of a national aerospace technology core curriculum.
- Promoted awareness and recruitment through the Air Force Career Field Manager, AFJROTC and Civil Air Patrol.
- Provided guidance in national skills assessment and technical professional development.
- Shared national certification knowledge, experience and skill requirements.
- Provided required curriculum database inputs.

Lessons Learned
The Air Force is highly interested in nationally recognized professional certifications. Its vision is to assist professional development by broadening the skills of its highly trained and qualified technicians. Blending industry-based education and providing certification opportunities will benefit the Air Force by molding a more diverse and qualified technician. The technician will

benefit by being provided the education and credential needed by industry when they transition from the Air Force. Likewise, industry will benefit immensely by receiving highly trained, qualified, experienced and disciplined technicians – a valuable payback. To capture our vision, intense networking is required to break down the barriers of military vs. industry standards, experience and skills. With SpaceTEC®, the organizational partnership resulted in solid leadership with innovative visions and persistence required to meet its goals.

Future Activities
CCAF will

- Develop a SpaceTEC® page on the CCAF website to educate and provide academic/certification guidance to Air Force technicians.
- Train and certify qualified Air Force technicians as SpaceTEC® Evaluators, empowering them to administer the Oral and Practical phase of the certification process. SpaceTEC® provided an Evaluator Kit to the Air Force that will be located at Vandenberg AFB, CA.
- Conduct a SpaceTEC® publicity campaign by attending and briefing Missile and Space Maintenance Utilization and Training Workshops, conferences and related forums.
- Attempt to develop an Air Force video advertising the SpaceTEC® initiative and demonstrating how it will benefit the Air Force and the aerospace industry.

President
Colonel Eric Ash
130 W Maxwell Blvd.
Maxwell AFB, AL 36112-6613

Vice President
Lt. Col. Michael Masterson

Co-PI/Contact
J.R. Breeding
Chief, Licensure/Certification Prog.
CCAF/DFAL
Maxwell AFB, AL 36112-6613
334-953-8423
j.r.breeding@maxwell.af.mil

Cuyahoga Community College Cleveland, OH

Background

Opened in Cleveland in 1963, Cuyahoga Community College is Ohio's first community college and Ohio's largest community college, serving approximately 60,000 students each year. More than 700,000 county residents have come through Tri-C's doors during the past four decades.

The College offers two-year associate degrees, certificate programs, and the first two years of a baccalaureate degree. The curriculum includes 900 credit courses in 70 career and technical programs and the liberal arts. Courses are offered at three Cuyahoga County campus locations and numerous off-campus sites. Tri-C's student teacher ratio is 19 to 1 and tuition is among the lowest in the state and the lowest in northeast Ohio. Thirty-six percent (36%) of our students take courses to prepare for transfer to a four-year institution, 61% are enrolled in technical job training courses, 66% are women, and 81% find jobs right here in Cuyahoga County.

Located in the Cleveland area is NASA's Glenn Research Center. The facility is noted for its high speed wind tunnel. Cleveland with a strong manufacturing base fits well into the Glenn Research Center developing fabrication and production techniques that enable aircraft designs to keep pace with the needs of a rapidly changing world. The skilled trades that maintain this facility and build the components that are tested at this NASA facility are trained at CCC. Companies such as Alcoa, Parker Hannifin, BF Goodrich, Keithly Instrument, and Swagelok rely on CCC to supply technicians, machinists, and fabricators to produce aerospace components.

Participation

The strength of Cuyahoga Community College (CCC) would be our connections to manufacturing and the aerospace industry. The Unified Technologies Center (UTC) has a manufacturing center that focuses on education and training in applied manufacturing and technology. The manufacturing center delivers education in

**The Manufacturing Center is housed at Unified Technologies
Center on the Metropolitan Campus of Cuyahoga Community College.**

fundamental and advanced CNC Technologies, Rapid Prototyping,
Computer Assisted Design, Electrical Systems and Instrumentation,
Fluid Power, Mechanical Systems, and Integrated Systems
Technology (IST).   With the changes and the advancement in
technology, CCC has adopted an integrated approach to all training.
We feel this fits well with idea of technicians needing to understand
complete systems. CCC would like to achieve:

- Development of core curriculum with pathways into different
  areas of aerospace.
- Better communication between industry and K-12 education.
  CCC has started inviting the local high schools to advisory
  meetings and industry based functions.
- Staff development programs that bring high school teachers to
  CCC making them aware of the available teaching resources.
- More marketing of SpaceTEC® and technology based careers.

Cuyahoga Community College has designed a strategy to
accomplish our goals and communicate the need for aerospace
technicians to the high school students by:

Developing a core curriculum for aerospace workers and high-tech
maintenance workers. The core curriculum has been developed with
the SpaceTEC® partners and the input of local companies such as

BF Goodrich, Swagelok, and Parker Hannafin. Along with these partners, CCC has asked the Cleveland Engineering Society to give us input and to help communicate the need for local aerospace engineers and companies to guide us as we develop the curriculum.

Hosting Mini-bots program that exposes area high school students and teachers to science and math applications (the theme of this year's mini-bots program was Mars exploration). This program brings the local area high school students together to build mini-bots from kits. The schools then compete on an obstacle course for trophies and shirts.

Developing and implementing survey classes demonstrating the applications of electrical, mechanical, fluid systems, and manufacturing processes to high school students and teachers. The high school students will have many hands-on exercises that teach the principles behind the mini-bots.

The start of one of the Mini-Bot Competition races. Students who have built their own robot cars, must maneuver them through the various obstacles of this course located on the Manufacturing Center Floor.

As part of a $1M grant awarded to Cuyahoga by the Department of Labor, this Amatrol equipment is one of many state of the art trainers used in various Industrial Maintenance and Integrated Systems Technology programs.

A state-of-the-art PLC (Programmable Logic Controller) Lab is located inside the Manufacturing Center.

Results

- This last year 17 Cleveland high schools participated in our Mini-bots program.
- 128 tenth and eleventh graders and teachers were exposed to engineering and technology.
- 93 students trained in Programmable Logic Controllers (PLC).
- 225 students trained in fundamental and advanced CNC technology.

- Many local industries were exposed to SpaceTEC® initiatives.
- A renewed bond has been developed between CCC and the NASA Glenn Research Center.
- 44 students completed Integrated Systems Technology (IST).
- The development of core curriculum and pathways into aerospace and high tech maintenance were completed.

Lessons Learned
The collaboration that SpaceTEC® fosters has opened lines of communication with industrial partners and NASA that have been missing from the Cleveland area. Also the renewed excitement of aerospace technology has local high schools and their teachers asking for more information. The grant has been an excellent source of networking providing CCC with additional opportunities to open dialogue with industry partners. CCC has benefited from SpaceTEC® by opening our eyes to core curriculum development and verifying that integrated training and a systems approach is what is needed in today's workplace.

Future Activities
Survey classes to supplement our high school mini-bots program will be implemented. These will teach students the application of the systems that make technology run. Also, the continued drive toward additional CNC technology and automation in manufacturing will prepare our students for the future manufacturing.

CCC has learned that the development of core curriculum with pathways to specific technology areas will help build a well rounded workforce. This will be accomplished by the implementation of 1 + 1 programs in Integrated Systems and Manufacturing training.

The SpaceTEC® program has been the perfect vehicle to reignite the educational fire for our youth in science and math. CCC has and will use SpaceTEC® to bring more K-12 schools to the college and help children understand what technology is and how it works. With the development of a certification and standardization of curriculum, aerospace has come back to life in the Cleveland area.

Kenneth McCreight, Program Manager for Machining Technology, gets students ready for the Mini-Bots competition.

President
Cuyahoga Community College District
Dr. Jerry Sue Thornton
700 Carnegie Avenue
Cleveland, OH 44115-2878
216-987-4854

Executive Vice President
Dr. Cullen Johnson

Co-PI/Contact
Craig McAtee
Exec. Dir., of
Manufacturing, UTC
2415 Woodland Avenue
Cleveland, OH 44115
216-987-3048
Craig.McAtee@tri-c.edu

Team Members
Dr. Charles George
Kenneth McCreight

Embry-Riddle Aeronautical University
Daytona Beach, FL
Prescott, AZ

Background

Embry-Riddle Aeronautical University teaches the science, practice, and business of the world of aviation and aerospace. Since its founding just 22 years after the Wright brothers' first flight, ERAU and its graduates have built an enviable record of achievement in aviation and aerospace.

ERAU offers more than 35 degree programs at residential campuses in Daytona Beach, FL, and Prescott, AZ, and at more than 130 Extended Campus Learning Centers for working professionals located throughout the United States and Europe. ERAU also provides distance education or online learning around the globe.

The curriculum at Embry-Riddle covers the operation, engineering, research, manufacturing, marketing, and management of modern aircraft and the systems that support them.

Embry-Riddle Aeronautical University Daytona Beach Campus

Enrollment for all campuses is an unduplicated annual headcount of 28,453 (2003).

Embry-Riddle students working in laboratory

In addition to its degree programs, ERAU, in partnership with airlines and corporations, provides non-degree customized flight training through the Commercial Airline Pilot Training program, which qualifies candidates for first-officer positions in transport-category aircraft. ERAU also offers training for aviation maintenance professionals through the Aviation Maintenance Technology (AMT) department. These programs include FAA-approved airframe and powerplant mechanic certification and a specialization in avionics line maintenance. Federal Communications Commission certification training is also available through AMT.

Participation
ERAU has worked with SpaceTEC® to secure articulation agreements with member colleges. In addition, ERAU hosted a visit by SpaceTEC® to their Daytona Beach campus in February 2004 to see the latest in aviation and aerospace technologies, and to learn about the programmatic options available.

Students in Airframe and. Powerplant Lab.

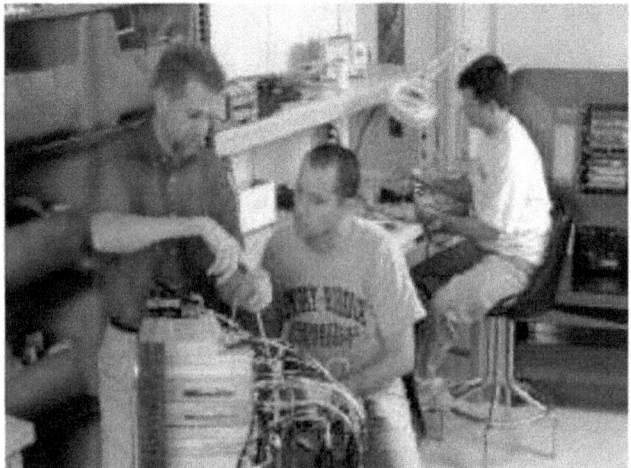

ERAU Instructor works with students in computerized laboratory.

ERAU plans to support articulation agreements that will allow community college students to enroll in ERAU programs to obtain their baccalaureate degree. Reverse-transfer agreements will allow ERAU AMT/ASM students to get college credit through one of the SpaceTEC® community college partners.

Results

Meetings have already occurred with many of the SpaceTEC® colleges to begin articulation discussions. Several of these articulation agreements have been finalized for signature and are in effect.

ERAU staff have partnered with SpaceTEC® representatives to expand aerospace marketing activities to key meetings including the Civil Air Patrol National Conference in Atlanta, GA, and the Experimental Aircraft Association Annual Experimental Air Show in Oshkosh, WI.

ERAU hosted a co-PI visit to its Daytona Facility to showcase the extensive new simulation technology and weather observation and forecasting capabilities for the future.

Lessons Learned

- The collaboration with SpaceTEC® benefits community colleges and their students by ensuring that potential transfer students and their faculty will be aware of the opportunities available and of the expectations of receiving institutions.

- ERAU will benefit from sharing years of experience in aviation and aerospace with SpaceTEC® faculty.

- Networking has increased our contacts and information about potential transfer students.

Future Activities

One key area where ERAU's NASA connection will assist SpaceTEC® work is the "Teach Space" project. This program hosts K-12 teachers to intensive training sessions that provide orientation to space-related curricula so that classrooms can be enriched in aerospace knowledge and terminology.

ERAU will become one of SpaceTEC®'s national certification testing centers. At least one ERAU instructor will be qualified as a SpaceTEC® Examiner (STE) and the college will operate the Test Center at its Daytona Campus using a SpaceTEC® Examination Kit.

President
Dr. George H. Ebbs
600 South Clyde Morris Blvd.
Daytona Beach, FL 32114

Provost, Chief Academic Officer
Dr. John P. Johnson

Chancellor, Daytona Beach Campus
Dr. Irwin Price

Chancellor, Extended Campus
Dr. Robert E. Myers

Co-PI/Contact
Dr. David Hosley
Dean of School of Corporate
Training and Professional
Development
386-323-8095
david.hosley@erau.edu

Team Members
Janet Dunn
Fred Mirgle

Palm Beach Community College
Lake Worth, Florida

Background

Palm Beach Community College is Florida's first public community college opening its doors in 1933. It has over 90 programs of study. It is 10th in the U.S. in the number of AA degrees awarded, 17th in total associate degrees awarded, and 34th in associate degrees awarded to minorities. The total enrollment is approximately 48,000 students both full and part-time. The average age of the student is 30. The college has five locations covering almost the entire Palm Beach County area with the main campus being the Lake Worth Campus.

Lake Worth campus of Palm Beach Community College.

PBCC does not presently have any arrangements with NASA, DOE, or DOD, but its proximity to Pratt-Whitney, a major rocket engine manufacturer, is important to this project. SpaceTEC® was the first Space-related partnership experienced by the college.

Participation

Initial efforts at PBCC were concentrated in outreach activities and providing research and recommendations regarding certification partners, curriculum development and national certification issues. Our approach was and is to create more alliances with the K-12 county school board, business development groups, and government groups and then gradually introduce SpaceTEC® topics, courses, and eventually the entire core curriculum to the college.

The start up of a massive state-supported workforce development program resulted in building and staffing a $20M building specifically built for workforce activities. The plan is to build on this wonderful workforce initiative by adding SpaceTEC® courses and topics wherever and whenever possible.

Results

In the first two years, outreach activities have included: K-12 teacher "Lunch and Launches" rocket workshops, on-campus rocket workshops, two board presentations, engineering class clean room exercises, a 30 student class participation with the Florida Space Research Institute Advanced Learning Environment (FSRI ALE) program, two hot rocket activities with the YMCA, a pilot project to introduce 8th through 12th graders to the Kennedy Space Center via bus, seven teachers sent to SpaceTEC® workshops, and an articulation agreement with Embry-Riddle University for the PBCC aviation program.

Lessons Learned

SpaceTEC® activities require the integration of the college administration with the faculty to access the working of the entire college.

Teachers at Lunch and Launch Workshop demonstrate Action vs. Reaction.

The "Lunch and Launch" activity requires careful planning, precise execution and a close relationship with the school board. If one of these items is missing, success is doubtful.

Adding courses and curricula to existing programs is a time consuming process, requiring a careful balance of persistence without overt pressure.

Most importantly, PBCC has created a foundation of trust and credibility with the local School Board, the business community, and a niche in the college technician and workforce initiatives on which to build future activities.

Contacts with local industry groups, government entities and other external groups are essential in obtaining college support and feedback to make any program move forward.

The networking of the SpaceTEC® Co-PIs is a vital element in successful exchange of information, introduction of new ideas, and reinforcing strategies and goals.

Future Activities

To maintain the SpaceTEC® momentum, the ever popular "Lunch and Launch" workshops and related activities for K-12 teachers will continue. College instructors will be sponsored to attend SpaceTEC® workshops and new alliances with business and industry will be formed.

Elements of the SpaceTEC® core curriculum will be introduced into the college and PBCC will sponsor five Team America Rocket Challenge (TARC) teams supplied by the local school board. PBCC professors will serve as advisors for these teams.

Engineering class doing clean room exercise.

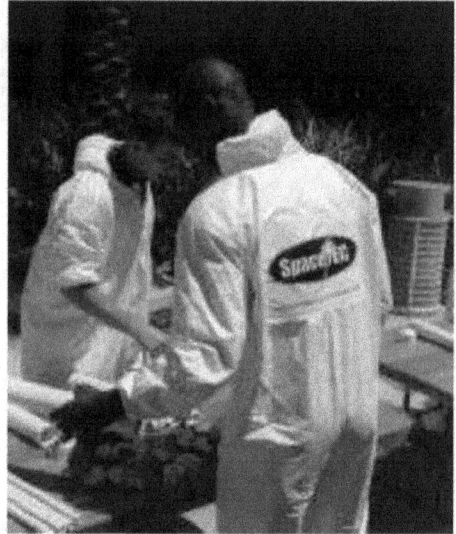

President
Dr. Dennis Gallon
Palm Beach Community College
4200 Congress Avenue
Lake Worth, FL33461-4796
561-868-3500

Vice President for Academics
Dr. Sharon Sass

Co-PI/Contact
Dr. Thomas Steffen
Professor III
Palm Beach
 Community College
4200 Congress Avenue
Lake Worth, FL 33461-
561-868-3417
steffent@pbcc.edu

Team Members
Fred Barch
John Berryman
Michael Calloway
Dorothy Rajcoomar
William Reddy
Jon Saken

Pearl River Community College Poplarville, MS

Background

Organized in 1909 as Pearl River County Agricultural High School, Pearl River Community College is among the oldest colleges of its kind in the South and is the pioneer junior college in the state. It has been the pathfinder for advanced education in southern Mississippi.

Located in Poplarville, Mississippi, Pearl River Community College is an open-admission, community based, comprehensive institution designed to provide economical and quality educational opportunities to residents of Forrest, Hancock, Jefferson Davis, Lamar, Marion, and Pearl River counties. The College, organized by the Board of Supervisors of Pearl River County and approved by the Legislature, has developed an educational mission characterized by diversification, growth, and community orientation. The mission of Pearl River Community College reflects this philosophy.

Pearl River Community College in Poplarville, Mississippi.

Pearl River Community College supports tenants and sub-contractors of NASA Stennis Space Center with a variety of academic and technical programs including metrology and GIS/GPS. The college offers over twenty-five degree and certificate programs in applied technology.

Participation
Pearl River Community College was an original member of the Community Colleges for Innovative Technology Transfer and was one of the nine original partner colleges in SpaceTEC®. PRCC staff have served as Chairman of the K-12 Program Delivery Coordination and represented SpaceTEC® at the National Tech-Prep Conferences.

Results
In addition to hosting annual coordinating conferences for K-12 and university representatives in Mississippi, PRCC has:

- Provided CX-Online licenses (Interactive Career Exploration software) for all SpaceTEC® partners.

- Hosted workshops for K-12 teachers, community college and university faculty to explore outreach and articulation options and resources for space-related curricula.

- Produced CDs for distribution to K-12 programs nationally, offering a directory of NASA Education Resources, a directory of Spatial Technology Programs including a glossary of associated terminology, and a metrology curriculum.

- Hosted SpaceTEC® workshops in "Metrology" and "How to be a DACUM facilitator" in 2004. The college has supported the SpaceTEC® national certification testing and will operate a testing center with a certified SpaceTEC® examiner.

Lessons Learned

Some key conclusions from PRCC's SpaceTEC® experience include:

- Partnerships are the key to success in applied technology programs. Pearl River has a strong partner in NASA Stennis Space Center and looks forward to continued growth there.

- The college continues to provide technical training in areas where over 400 students can find employment in the area.

- SpaceTEC® partner colleges have contributed to curriculum development and K-12 outreach through the sharing of existing materials.

- SpaceTEC® partners can aid each other in program development in these times of tight fiscal management.

Electronics Laboratory at Pearl River Community College.

Future Activities

Pearl River Community College has assigned SpaceTEC® activities for the future to new leadership in the Technical Education Division of the college. Near term activities will include:

- Support SpaceTEC® certification program, including evaluator appointment and training, and operation of a national core certification testing center including an examination kit and a certified SpaceTEC® Examiner (STE).

- Initiate announced SpaceTEC® related curriculum in fall of 2004.

- Utilize SpaceTEC® partners and networking to strengthen a new aviation maintenance program.

- Continue support to SpaceTEC® activities linking PRCC and Stennis Space Center where local needs have been identified

President
Dr. William A. Lewis
101 Highway 11 North
Poplarville, MS 39470-5060

Vice President for Instruction
Dr. John Grant
jgrant@prcc.edu

Director, Career Technical Education
Mr. Don Welsh

Co-PI/
Contact
Ms. Elaine Smith
Asst. Dir., Career
Technical Ed.
601-403-1240
elsmith@prcc.edu

Team Member
Victor Cerniglia

Prince George's Community College Largo, MD

Background

Prince George's Community College, located in suburban Washington, D.C., was a member of the first Community Colleges for Innovative Technology Transfer (CCITT) national NSF grant. The college enrolls approximately 13,000 credit students each semester. In addition, over 20,000 students take non-credit courses and workshops annually.

Prince George's Community College (PGCC) student population is 85% minority, with African Americans representing the largest minority. The college's main campus is in Largo, just outside the Capital Beltway.

Other College Centers are located at Andrews Air Force Base, the University Town Center in northern Prince George's County, and the new Laurel College Center, a joint initiative between PGCC and neighboring Howard Community College. With NASA Goddard Space Flight Center located in Prince George's County, the college has a strong relationship with Goddard and its contractors.

PGCC comes to SpaceTEC® with a program in Space Technology modeled after that at Brevard Community College. The Space Technology Program is one of four options within the Engineering Technology Program.

PGCC is responsive to local needs. The college has linked with neighboring Capitol College, Honeywell, and NASA Goddard to offer PGCC engineering and engineering technology students internships in the satellite monitoring facility at NASA Goddard.

This program is designed to produce students who will complete a four year B.S. degree in astronautical engineering. To date, three PGCC students have had paid internships at the satellite monitoring facility at NASA Goddard.

Participation
Prince George's County has limited manufacturing. NASA Goddard's research mission involves satellite tracking, support of the Hubble Space Telescope, and earth system science.

Through SpaceTEC®, the college has assessed its Space Technology and Engineering Technology offerings, strengthened the interaction between engineering faculty and engineering technology faculty, leveraged equipment and software, increased its recruitment into area high schools, and developed a collaborative relationship with Laurel High School which has a specialized satellite program.

Engineering Technology faculty member Charles Hendrickson works with students on wireless technology.

Some of the engineering technology courses have been formatted for distance education to meet the needs of our students. SpaceTEC® also served as a co-sponsor for several science, engineering, technology and mathematics student academic awards programs.

Results

In an attempt to strengthen recruitment into Space Technology and Engineering Technology, SpaceTEC® sponsored its first Space Scholars Institute for 10th through 12th grade students in summer 2004. Recruitment was on a competitive basis and required specific

Students in the Space Scholars Institute studied the history of space vehicles. In the lab, they put together small Trebuchets and studied how different variables impacted on the projectile distance.

grades, course work, and a teacher letter of recommendation. The curriculum for the two-week Institute was hands-on and included an introduction to soldering, work on circuit boards, the making of an electronic "space bug", and the design of a robotic Mars Rover. The program culminated in team research presentations at an evening event attended by parents, teachers, and friends.

Students and faculty also toured the Johns Hopkins University's Applied Physics Laboratory in Columbia, MD, and the NASA Goddard Space Flight Center during the Institute.

The Institute received excellent publicity in local newspapers and on the local Cable TV station. Space Scholars were interviewed on several occasions during their two weeks at the college.

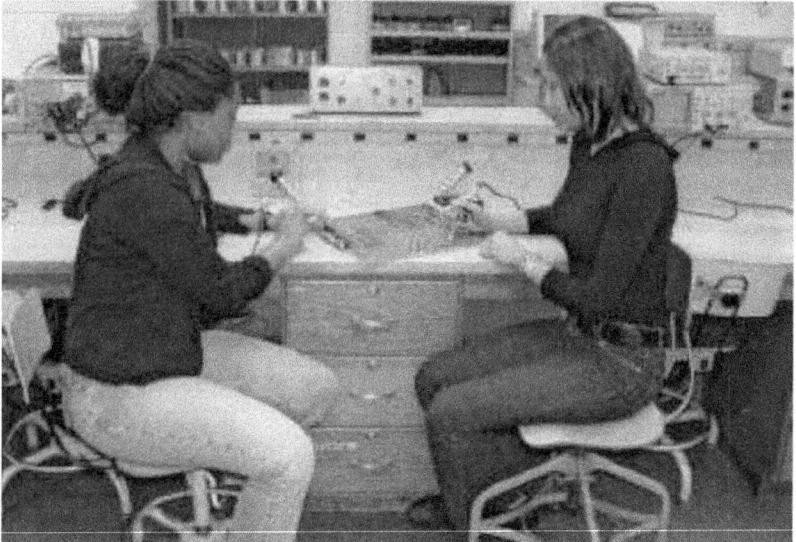

Two Space Scholars solder a circuit board.

The college also hosted a SpaceTEC® Faculty Professional Development workshop in April 2003.

The college played a lead role in the development of the SpaceTEC® Monograph which highlights the accomplishments of this National NSF Center of Excellence as well those of the individual colleges. This publication will be disseminated broadly to industry, government, and other academic institutions. The college also serves frequently as a representative for SpaceTEC® at meetings and conferences in the Washington area.

Articulation of the college's engineering technology options, which include Space Technology, is underway with Embry-Riddle

Aeronautical University. Other articulation agreements exist with Capitol College in Laurel, MD.

Lessons Learned

Local business and industry drive the technology curricula in community colleges. In order for community colleges to meet these needs and to take advantage of them when they arise, the colleges must be engaged in the broader community on an ongoing basis.

Other SpaceTEC® colleges have built strong industry involvement that has helped them build their programs around needs specific to manufacturing in their local areas. Even though there are limited manufacturing jobs in the Washington area, the basic processes involved in manufacturing, fabrication, and composites are still highly relevant to the curriculum at our (and any other) college. It is therefore possible to learn how these processes work and mold them to local needs.

Future Activities

- SpaceTEC® will be a co-sponsor of Science Trek, a fall Conference for 250 4th-6th grade students and their parents to introduce them to all areas of science and technology.
- Additional online course development for engineering technology is planned.
- SpaceTEC® will be a co-sponsor for student tours to industry through the newly established STEM (Science, Technology, Engineering, Mathematics) Collegian Center at the college.
- Efforts are underway to build on the strong relationships that were developed between the college, high school principals, high school teachers and the Summer 2004 Space Scholars.
- Specialized programs are planned for the Space Scholars and other interested high school students for fall 2004.
- PGCC's SpaceTEC® will become involved in the certification process through the grant.
- SpaceTEC® will continue to seek additional internships and collaborations with NASA Goddard and its contractors.
- A second Space Scholars Institute will be held in summer 2005.

A Space Scholar ponders design options for his
Mars Rover.

President
Dr. Ronald A. Williams
Prince George's Community College
301 Largo Road
Largo, MD 20774-2199
301-336-6000

Vice President for Instruction
Dr. Vera Zdravkovich

**Dean of Sciences, Technology,
 Engineering, and Mathematics**
Dr. Aaron Stucker

Co-PI/Contact
Dr. Patricia A. Cunniff
Professor Emeritus
and Former Dean
Prince George's
 Community College
pcunniff@pgcc.edu

Team Members
Jack Bailey
Robert Bard
Charles Hendrickson
William Lauffer
Allison Miner
Leslie Wojciechowicz

San Jacinto College District Pasadena, TX

Background

San Jacinto College District is a comprehensive, public, community college that blends traditional methodologies and technology-based delivery systems.   The college delivers accessible, affordable, high-quality basic skills, academic support, and post-secondary education programs, which prepare students to enter the job market,

San Jacinto College District has recently completed construction of an Interactive Learning Center on each of its three campuses. The Centers include wireless computer classrooms and conference rooms as well as an open wireless computer lab for student use.

transfer to senior institutions, achieve their personal and professional goals, and assume leadership roles in a culturally and ethnically diverse global community. Through its programs, services and partnerships with industry, the college supports the economic growth of the community and the region.

San Jacinto College has been providing workers for the aerospace industry since the early 1960s when both the college and the Johnson Space Center were established. In 2000 San Jacinto's relationship with aerospace was formalized with the creation of the Aerospace Academy (housed at the Johnson Space Center), through which SpaceTEC® and other aerospace-related activities are coordinated. The Aerospace Academy, a K-20 education-industry-government collaboration, includes NASA-Johnson Space Center and its area employers, Space Center Houston, Bay Area Houston Economic Partnership, three universities, and four school districts along with San Jacinto College.

Demonstrating the value the college brings to the aerospace industry are two points:
- Many of the 184,175 aerospace and aviation employees in Texas received their initial education at San Jacinto College.
- Since 1996 almost 15,000 incumbent workers in 43 Johnson Space Flight Center-area organizations have received free upgrade training through $8M in grants that San Jacinto has acquired for the industry.

Participation
San Jacinto College's SpaceTEC® goals include the following:
- Implement retention and outreach through promoting SpaceTEC®/NSFATE.
- Identify local concentrations of SpaceTEC® curricula.
- Develop K-12 programs to engage students in science, technology, engineering, and math.
- Produce Concentration Certification Exams and focus relevant San Jacinto College programs to SpaceTEC® core curriculum.
- Distribute Learning Modules to K-12 educators.

- Produce a coordinating tutorial discussing the use of the Modules for professional development.

San Jacinto College is allied with the SpaceTEC® curriculum primarily through three associate degree programs: Geographic Information Systems (GIS), Aeronautical Technology, and Pre-Engineering.

For individuals seeking careers in aviation, San Jacinto College Central has developed four separate programs, all leading to an Associate of Applied Science degree or certificate in Aeronautical Technology.

Results
Through linking and leveraging of SpaceTEC® and related programs/grants during the last two years, San Jacinto College has achieved and/or is pursuing the following:

- Received a grant from the Texas Space Grant Consortium to extend GIS training to 25 teachers and provide them with the training and resources needed to integrate science and space concepts and NASA resources into lesson plans tied to state and national science standards.
- Received a National Science Foundation grant to provide scholarships to computer science, engineering, and mathematics majors, particularly targeting underrepresented and economically disadvantaged students. The Academy is

partnering with University of Houston-Clear Lake to provide additional scholarships.

- Partnered with SpaceTEC® member Embry-Riddle Aeronautical University on articulation and other activities to further SpaceTEC® goals.
- Created a multi-subject lesson module for K-12 educators to help promote interest in aerospace careers in K-12 students. The *SpaceTEC® Science-Technology-Math Lessons: Resources for K-12 Educators CD* was field tested at the Alabama Aerospace Educators conference in June 2004 in conjunction with SpaceTEC® partner Calhoun Community College and received excellent feedback and recommendations for wide dissemination.
- Helped sponsor the 2004 NASA's Educator Astronaut Program and their training, focused on "Inspiring the Next Generation of Explorers." These 160 science and math educators from around the nation are part of Network of Educator Astronaut Teachers (NEAT).

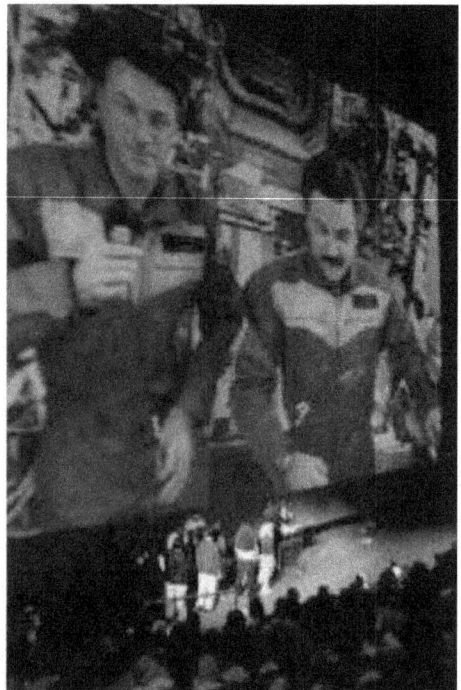

Astronaut Michael Foale and Cosmonaut Alexander Kaleri answered questions from Houston area middle school students live from the International Space Station in a SpaceTEC® sponsored activity.

- Attended "GIS Day Conference" at University of Houston-Downtown and "Conference for the Advancement of Science Teachers" (CAST), with an outreach of over 6,000 for both conferences. Hosted a three-day career fest at Space Center Houston where 75 middle and high school instructors heard several area aerospace industry employees from companies such as NASA, Lockheed Martin, and Boeing speak.
- With NASA-JSC and Space Center Houston, provided middle school science teacher training supplemented by a "Teaching from Space" downlink for 900 students with astronauts on the International Space Station answering student questions in real time.
- Coordinated the "Making Connections" workshop for SpaceTEC® partners.
- Trained 954 additional math/science teachers in 66 school districts, community colleges and universities across Texas under grants. Additional training now occurring under grants.
- Trained 3,703 current and new aerospace employees for 41 Johnson Space Center-area employers under grants.
- Coordinating NASA-JSC's KC-135 Student Flight Program/Graduate Student Research Program/NASA Faculty Fellowship Program Administration and the Community College Aerospace Scholars Program.
- Presentations highlighting SpaceTEC® have been made at the League for Innovation conference, Texas Community College Teachers Association conference, and Texas Workforce Commission conference.

Lessons Learned

Responding to the increasing workforce needs of the aerospace industry creates a challenge for individual educational institutions. Of all of the lessons learned from participating in the SpaceTEC® program, one of the most important is the benefit that can come from linking and leveraging. The strength of an organization or institution lies in its ability to form supportive, enduring partnerships in both the community and in industry. An effective partnership is one through which the strengths of individual partners provide a mutual benefit to all parties involved. The creation and achievements of the SpaceTEC® network underscore the benefits to be derived from collaboration. Linking and leveraging enables member schools and their local K-12 and university partners to use the strengths of all partners, share products and helpful experiences, and to piggy-back rather than unnecessarily duplicating work or starting from zero.

Future Activities
1. Complete the delivery of five modules providing topical aerospace educational materials in CD format to all partner colleges with a coordinated tutorial on their use for professional development.
2. Conduct an internal faculty workshop for applied technology at San Jacinto College with partial funding by SpaceTEC® with feedback indicating impacts in support of NSFATE goals.
3. Provide assistance to the aviation technology program at San Jacinto College in support of the SpaceTEC® core curriculum beginning fall 2004.
4. Bio Technology and Bio Science curricula are currently under development and will adhere to the SpaceTEC® CORE curriculum.
5. Continue to address aerospace workforce and skills shortages.
6. Continue creating highly qualified science/mathematics teachers to produce the needed workforce.

Chancellor
San Jacinto College District
Dr. Bill Lindemann
4624 Fairmont Pkwy.
Pasadena, TX 77504
281-998-6100

Vice Chancellor
Dr. Gwen Tilley
4624 Fairmont Pkwy.
Pasadena, TX 7504
281-998-6182

Co-PI/Contact
Dr. Marie Dalton
Executive Vice President
Aerospace Academy
2101 NASA Pkwy. Bldg 226
Mail Code AHS
Houston, TX 77058
281-483-1145
marie.dalton1@jsc.nasa.gov

SpaceTEC® Project Coordinator
Christy Mapes
281-483-1146

Thomas Nelson Community College Hampton, VA

Background

Thomas Nelson Community College (TNCC) is a comprehensive community college located in historic southeast Virginia. The college serves the cities of Hampton, Newport News, Poquoson, and Williamsburg and the counties of York and James City. The College had a Fall 2004 enrollment of approximately 5300 FTE.

Thomas Nelson Community College Campus.

The community college also serves numerous Air Force, Army, Navy, and Coast Guard installations. Langley Air Force Base is located just minutes from the campus and is adjacent to a NASA research center. The largest local industry is Northrop Grumman Newport News which has contracts for designing, building and overhauling U.S. Navy aircraft carriers and submarines.

TNCC is working in partnership with NASA and other industry and educational leaders to develop a SpaceTEC® Advisory Council whose vision is to develop a new generation of technologists for the area. Membership on the council will include: NASA Langley, Northrop Grumman, Allied Aerospace, Advanced Technologies, the Space Grant Consortium, Virginia Air and Space Center, Old Dominion University, Norfolk State University, the Peninsula Alliance for Economic Development, and the five Peninsula school districts. Thomas Nelson Community College will provide the administrative leadership for this group.

Participation
In recent months, TNCC has been working with a team of NASA engineers to determine skills expected of new graduates in Engineering Technology. As a result, TNCC full-time and adjunct faculty are now developing five courses of study for the purpose of standing up the SpaceTEC® core curriculum as the foundation of new degree programs. Two currently offered TNCC courses are being modified to incorporate all objectives of the Tests and Measurements and Applied Mechanics SpaceTEC® courses into equivalent TNCC courses. Three SpaceTEC® courses, Introduction to Aerospace Technology and Materials Processes I and II, are new to TNCC and to the Virginia Community College System and are requiring more extensive curriculum development. All five core curriculum SpaceTEC® courses will be the foundational courses in the new AAS degrees in Technical Studies with specializations offered in Engineering Technology and Materials Science. Each of these degrees requires six credits of work-based learning.

TNCC student working in computer lab.

The college is also partnering with the Virginia Space Grant Consortium to develop an Introduction to Aerospace curricula that will be offered both at the community college and, through dual-enrollment, to partnering school divisions. Additionally, a team of aerospace and materials science engineers is currently working with TNCC administration and faculty to develop courses in Materials Science, Metrology, and Applied Mechanics as the core of two future SpaceTEC® associate degree programs. One of these degree programs will specialize in materials science and will constitute the first Materials Science degree program in the Virginia Community College System. The second degree program will specialize in Engineering Technology.

Results
In the six months since committing to the SpaceTEC® consortium, TNCC has accomplished the following:

- Identified courses for two new degree programs and related career studies certificates.
- Established teams of faculty and industry representatives to develop SpaceTEC® core curriculum courses.
- Partnered with area high schools to identify curricula programs that will prepare students to succeed in new SpaceTEC® degree programs. Assisted these schools in obtaining funding to provide teacher and counselor training and equipment and supplies to launch Engineering Technology curricula in fall 2004.
- Established the SpaceTEC® Advisory Council and developed a vision, mission, values and strategic plan for the new business-education partnership.
- Trained faculty in Materials Science and as DACUM facilitators through participation in SpaceTEC® professional development activities.
- Obtained media coverage for new SpaceTEC® partnership.
- Obtained funding and contracted for development of new materials science lab on campus.
- Partnered with Old Dominion University in submission of a NSF grant proposal targeted to increasing articulation between TNCC and ODU Engineering Technology programs.

- Established industry and community partnerships that will provide K-12 teacher and counselor training, work sitelearning for TNCC SpaceTEC® students, and additional Scholarships for TNCC students in SpaceTEC® programs.

Lessons Learned

The National SpaceTEC® Center and its consortium colleges have been a valuable resource to TNCC in the development of two new degree programs. NSF faculty development opportunities, such as the recent Materials Science workshop at Antelope Valley Community College, have particularly helped TNCC by sharing their curricula and information relevant to the design of the new materials science lab and procurement of equipment and supplies.

In its first six months of SpaceTEC® activity, lessons learned include the following:

- Community and industry partnerships are essential to delivery of program elements essential for student recruitment and retention.
- Developing and maintaining these partnerships requires substantial administrative commitment.
- Curriculum is improved when it is collaboratively developed by industry and educators.
- Keeping abreast of other NSF programs is essential to effective and efficient development of curriculum and extracurricular programs.
- Student recruitment begins with meaningful partnerships with area high schools to include secondary and post-secondary curriculum development.

Future Activities

A team of business and industry advisors, post-secondary faculty and secondary teachers will be formed through the Virginia Peninsula Tech Prep Consortium, in fall 2004, to identify and develop dual enrollment courses within the SpaceTEC® core curriculum. Additionally this group will explore incorporating the K-12 modules developed by SpaceTEC® partners into career education events targeted to secondary school teachers, counselors, parents, and students.

Current renovations at the college include design of a new Materials Science lab to be completed by Fall 2005. Utilization of this lab for the core courses in the SpaceTEC® curriculum requires the development of an equipment and supply list and acquisition of required items. The College has dedicated a portion of its Perkins funds for equipment purchases in this area and the college foundation is actively pursuing donations of equipment from business and industry for this purpose.

The TNCC SpaceTEC® Advisory Council envisions development of career pathways for future SpaceTEC® students that begin at the middle school level and continue through articulation into four-year university programs in Engineering Technology and Materials Science. Planning is currently underway to provide middle and high school teacher training, professional development opportunities for counselors, college scholarships, business and industry mentoring for students at all educational levels, and relevant dual-enrollment college courses in partnering high schools.

In addition to its two new degree programs, TNCC will offer career studies certificates that are comprised of the core SpaceTEC® courses. These will be marketed to students or technicians who already hold associate or bachelor degrees but wish to increase their knowledge of engineering technology in aerospace.

President
Dr. Charles Taylor
Thomas Nelson Community College
P.O. Box 9407
Hampton, VA 23670
757-825-2711

**Vice President for Academics
  And Student Affairs**
Dr. Ronald Davis

Co-PI/Contact
Patricia Taylor
Dean, Math, Engr.,
  and Technologies
757-825-2898
taylorp@tncc.vccs.edu

Team Member
M. Elizabeth Creamer
Nancy Holloway
Rudy Schwab

82

The SpaceTEC Program is already very broad
in scope, linking a network of government
agencies, business and industry groups, and
educational institutions. The consortium is
characterized by strong ties between local
and national organizations as shown here.

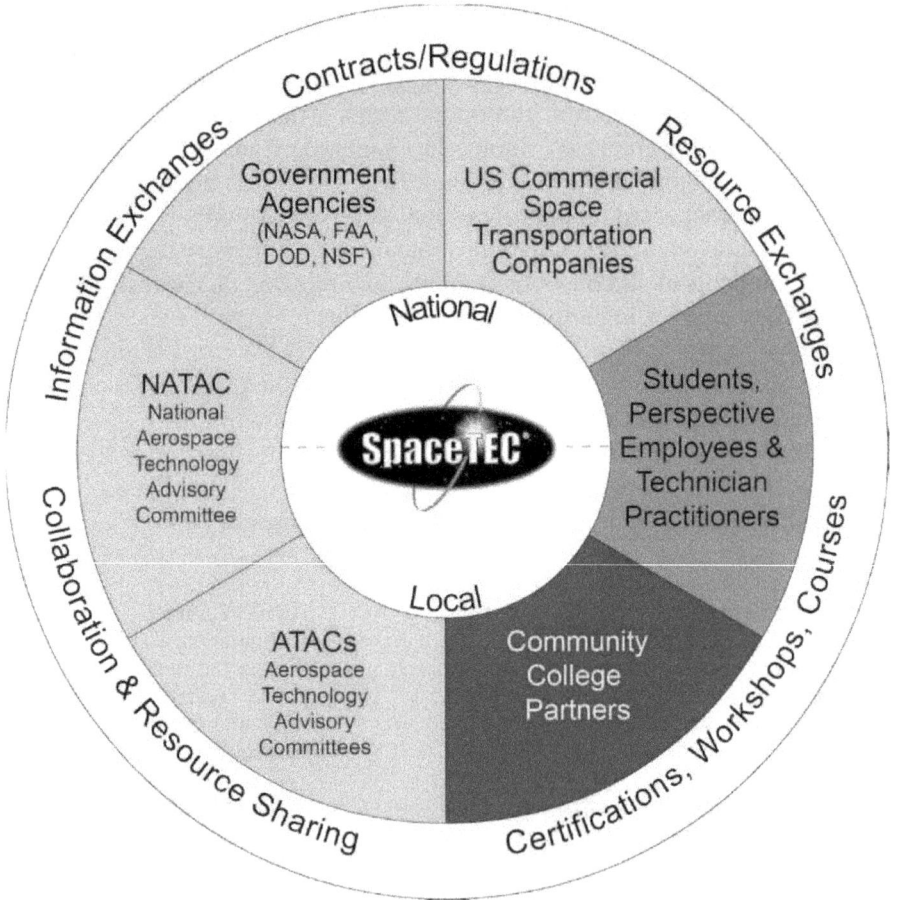

Challenges and Opportunities

Albert Koller, Principal Investigator, SpaceTEC®

The SpaceTEC® achievements catalogued in this monograph provide an important view of a unique and fascinating industry about which the average American knows very little. The demanding nature of the work masks some of the important lessons for broader applications. For example, many of the same issues now facing the aerospace industry – including aging workers and skills deficits – will soon arise in other technical areas. The Department of Labor's "President's High Growth Job Initiative" is addressing similar issues across twelve key industries nationwide. If America's technical workforce is to remain globally competitive, steps must be taken to remedy these shortcomings before they impact the economy. The lessons learned at SpaceTEC® provide many opportunities to reduce costs and avoid problems in the future.

Because much of our early education after the first year or two of schooling focuses on mental rather than physical skills, there is a tendency to ignore or de-emphasize higher-level motor training activities. This has resulted in a lack of good textbooks and related instructional materials for practical manipulative skills at the high school and post-secondary levels. SpaceTEC® partner colleges have taken steps to reduce or eliminate these shortcomings by developing curricula, text materials, lab manuals, and workshops that include creative hands-on exercises that demonstrate the importance of reasoning skills to success in performing physical work. Rocket workshops and field activities draw upon the competitive nature of Americans of all ages to make learning fun, often without exposure to any formal curriculum. These activities result in an infectious enthusiasm and a willingness to pursue new opportunities for active learning, including matriculation into formal degree programs.

An early result of the SpaceTEC® experience has been the transition of students who initially underachieve in mathematics and basic science to higher levels of interest and understanding of these essential areas. Much more work like this is possible in most of the relevant technical areas.

An important challenge that may not be obvious is the general social status of the American technical worker. To many Americans, those in the technical workforce are invisible; they simply do not exist in the mainstream of our society. At one end of the "respect spectrum" are athletes and entertainers; at the other end are the technicians, plumbers, auto mechanics, and others who work with their hands. The negative attitudes toward technical work expressed by our society discourage young people from entering careers in areas where hands-on skills are essential to success, even though technicians are typically paid at higher rates than their counterparts in service industries, as an example. SpaceTEC® colleges have set about addressing these shortcomings through specific partnerships and actions ranging from recruitment and marketing to sponsoring national competitions for youth. A good example is the sponsorship of "Technician Appreciation Night" at the annual "Space Congress" at Cape Canaveral, the first such event for "hands-on" technicians and mechanics in its 41-year history and enthusiastically endorsed by all.

As a result of the attitudes discussed above, our technical workforce is not recognized as a profession. There presently are no formal career ladders in aerospace and few plans exist for continuing education leading to degrees. SpaceTEC® sponsored databases for job banks, refresher training programs, and the implementation of a national certification program are aimed at improving the professional status of technicians nationwide.

With continued support by the American Technical Education Association and the National Science Foundation's ATE Program, SpaceTEC® holds at least some of the keys to championing the U. S. technical workforce of the future.

SpaceTEC® National Visiting Committee

Dr. Robert Sullivan, Chair
Vice President, Research & Graduate Progs.
Florida Institute of Technology
150 W. University Blvd.
Melbourne, FL 32901

Dr. Ron Bobay
Area IV Superintendent
Brevard County School Board
850 Knox McRae Drive
Titusville, FL 32780

Susan Moore
Director, People and Administration
Decatur Delta Operations,
The Boeing Company
100 Decatur Way
Trinity, AL 35673

Michael Flynn
Director, Apprenticeship & Training
International Association of Machinists
    and Aerospace Workers (IAMAW)
Upper Marlboro, MD 20772

Ed Gormel, Retired
Florida Space Authority
2727 Newfound Harbor Drive
Merritt Island, FL 32952

Dr. Maxwell King
President Emeritus
Brevard Comm. College
1384 Walton Heath Court
Rockledge, FL 32955

Jenifer Santer
Lockheed Martin
111 W. Betteravia
Santa Maria, CA 93455

Dr. Elane Seebo, Dean
Wichita Falls, TX
  Campuses Wayland
  Baptist University
426 5th Ave., Ste. 7
Sheppard AFB, TX

Mr. Lee Solid, Retired
Rocketdyne
765 River Oaks Lane
Merritt Island, FL 32953

Ms. Midge Davis, V.P.
Barrios Technologies
2525 Bay Area Blvd.
Ste. 300
Houston, TX 77058

Notes

National Aerospace Technology Advisory Committee (NATAC)

Mr. Marshall L. Heard, Chairman
Florida Aviation & Aerospace Alliance

Committee Members:
William Allred, Vice President
American Technical Educators Association (ATEA)

Garcia Blount
NASA, Goddard Flight Research Center
Greenbelt, MD 20771

Midge Davis, Vice President
Barrios Technology
Houston, TX 77058

Larry Gooch
Vice President & Chief Technical Officer
California Space Authority
Santa Maria, CA 93455

George Hauer, General Manager
Wyle Laboratories Florida Operations
Kennedy Space Center, FL 32899

Ann O. Heyward, Vice President
Workforce Enhancement
Ohio Aerospace Institute
Cleveland, OH 44142

Adrian Laffitte, Atlas Program Director
Lockheed Martin Launch Operations
Cape Canaveral Air Force Station

Jeff Little, Chief Technologist
J.E. Sverdrup
Huntsville, AL 35806

Susan Miller, Director
Office of Academic Investments
Dryden Flight Research Center
Edwards, CA 93523-0273

Geoff Schuler, Director, Operations
Missile Defense Systems
The Boeing Company
Huntsville, AL 35824-6402

Winston Scott, Director
Florida Space Authority
Cape Canaveral, Fl 32920

John Vickers, NASA
National Center for Advanced Manufacturing
Huntsville, AL 35812

Alfred Wassel
Commercial Space Transportation Safety Office, Federal Aviation
Association Patrick AFB, FL 32925-4355

Co-Principal Investigators

Allan Hancock College

Ardis Neilsen
Director of Economic Development
Allan Hancock College
800 South College Drive
Santa Maria, CA 93454-6399
805-922-6966 x3325
aneilsen@hancockcollege.edu

Antelope Valley College

Margaret Drake
Dean of Technical Education
Antelope Valley College
3041 West Avenue K
Lancaster, CA 93536-5426
661-722-6327
mdrake@avc.edu

Brevard Community
College

George Strohm
Associate Director
Aerospace Programs
Brevard Comm. College
BCC-A, M6-306, Rm.2000
Kennedy Space Ctr, Fl 32899
321-449-5060
strohmg@brevardcc.edu

Calhoun Community
College

James Swindell
Asst. Dean of Technology Education
Calhoun Comm. College
P.O. Box 2216
Decatur, AL 35609
256-306-2539
jes@calhoun.edu

Community College
of the Air Force

J. R. Breeding
Chief, Licensure/Certification Branch
CCAF/DFAL
Maxwell AFB, AL 36112-6613
334-953-8423
j.r.breeding@maxwell.af.mil

Cuyahoga Community
   College

Craig McAtee
Exec. Dir., Manufacturing
 and Applied Technologies
Cuyahoga Comm. College
2415 Woodland Avenue
Cleveland, OH 44115
216-987-3048
Craig.McAtee@tri-c.edu

Embry-Riddle
   Aeronautical University

Dr. David Hosley
Dean, School of Corporate
 Training and Professional
 Development
ERAU
600 South Clyde Morris Blvd.
Daytona Beach, FL 32114-3900
386-323-8095
david.hosley@erau.edu

Palm Beach Community
   College

Dr. Thomas Steffen
Professor III
Palm Beach Comm. College
4200 Congress Avenue
Lake Worth, FL 33461-4796
561-868-3417
steffent@pbcc.edu

Pearl River Community
   College

Elaine Smith
Asst. Director
Career Technical Education
Pearl River Comm. College
101 Highway 11 North
Poplarville, MS 39470-5060
601-403-1240
elsmith@prcc.edu

Prince George's Community College

Dr. Patricia A. Cunniff
Professor Emeritus
Prince George's Community
College
301 Largo Road
Largo, MD 20774-2199
301-322-0432
pcunniff@pgcc.edu

San Jacinto College

Dr. Marie Dalton
Exec. Vice President
Aerospace Academy
San Jacinto College
2101 NASA Parkway, Bld. 226
Mail Code AHS
Houston, TX 77058
281-483-1145
Marie.dalton1@jsc.nasa.gov

Thomas Nelson Community
  College

Patricia Taylor
Dean, Mathematics, Engineering
  and Technologies
Thomas Nelson Community
  College
90 Thomas Nelson Drive
P.O. Box 9407
Hampton, VA 23670
757-825-2898
taylorp@tncc.vccs.edu

Notes

National Center Staff

Dr. Al Koller, Principal Investigator
Executive Director, SpaceTEC®
SpaceTEC® Headquarters
MC: BCC/SpaceTEC®
Kennedy Space Center, FL 32899
321-730-1020
kollera@brevardcc.edu

David Brotemarkle
Project Manager, SpaceTEC®
321-730-1020
brotemarkled@brevardcc.edu

Juanita Curtis
Program Coordinator, SpaceTEC®
321-730-1020
curtisj@brevardcc.edu

Steve Pantano
Technical Assistant, SpaceTEC®
321-730-1020
pantanos@brevardcc.edu

Benny Dawes
Facility Manager, SpaceTEC®
321-730-1020
dawesb@brevardcc.edu

Notes

Notes

www.ingramcontent.com/pod-product-compliance
Lightning Source LLC
Chambersburg PA
CBHW060635210326

41520CB00010B/1616